Component-Based Systems

Component-Based Systems

Estimating Efforts Using Soft Computing Techniques

Kirti Seth, Ashish Seth, and
Aprna Tripathi

CRC Press
Taylor & Francis Group
Boca Raton London New York

CRC Press is an imprint of the
Taylor & Francis Group, an **informa** business

MATLAB® is a trademark of The MathWorks, Inc. and is used with permission. The MathWorks does not warrant the accuracy of the text or exercises in this book. This book's use or discussion of MATLAB® software or related products does not constitute endorsement or sponsorship by The MathWorks of a particular pedagogical approach or particular use of the MATLAB® software.

First edition published 2021
by CRC Press
6000 Broken Sound Parkway NW, Suite 300, Boca Raton, FL 33487-2742

and by CRC Press
2 Park Square, Milton Park, Abingdon, Oxon, OX14 4RN

© 2021 Taylor & Francis Group, LLC

CRC Press is an imprint of Taylor & Francis Group, LLC

Library of Congress Cataloging-in-Publication Data

Names: Seth, Kirti, 1966- author. | Seth, Ashish, author. | Tripathi, Aprna, author.
Title: Component-based systems : estimating efforts using soft computing techniques / Kirti Seth, Ashish Seth, and Aprna Tripathi.
Description: First edition. | Boca Raton : CRC Press, 2020. | Includes bibliographical references and index.
Identifiers: LCCN 2020022750 (print) | LCCN 2020022751 (ebook) | ISBN 9780367856090 (hbk) | ISBN 9780367441753 (pbk) | ISBN 9781003013884 (ebk)
Subjects: LCSH: Software engineering--Management. | Component software. | Soft computing.
Classification: LCC QA76.758 .S4583 2020 (print) | LCC QA76.758 (ebook) | DDC 005.068--dc23
LC record available at https://lccn.loc.gov/2020022750
LC ebook record available at https://lccn.loc.gov/2020022751

ISBN: 978-0-367-85609-0 (hbk)
ISBN: 978-0-367-44175-3 (pbk)
ISBN: 978-1-003-01388-4 (ebk)

Typeset in Times
by Deanta Global Publishing Services, Chennai, India

Contents

Preface

Software effort estimation is one of the oldest and most important problems in software project management. Today there is a vast array of models, each with its own unique strengths and weaknesses. Moreover, these models have also been proposed in relation to the environment and the context in which they can be applied. A new type of software systems known as Component-Based Systems have been trending for the last three to four decades. These systems make software development as simple as a plug and play device. This approach primarily focuses on abstraction and reusability.

For the processes such as project planning, project budgeting, and project bidding, the estimation of effort in software projects is considered to be fundamental. This effort estimation is taken seriously in software companies because improper planning and budgeting will lead to troubling consequences. So, effective estimation is necessary for software development.

In project management, the estimation of effort in software development has recently emerged. To date, in cases that consider prediction and estimation, there are still unresolved issues. In addition, if we adopt higher reliability then the process of effort estimation is not yet possible and is a challenge. Therefore, the process of estimating effort is still a tedious process for project managers.

Over the past few decades, effort estimations for software systems have caught the attention of researchers. Although there are techniques for effort estimation of legacy systems, there are very few techniques for estimating efforts in Component-Based Systems. This book focuses on the effort estimation of Component-Based Systems based on Soft Computing Techniques. Specifically, this book focuses on the following three techniques:

- Mathematical Model-Based Techniques.
- Fuzzy Logic-Based Techniques.
- Genetic Algorithm-Based Techniques.

This book will be very useful for students – both advanced undergraduates and entry-level graduate students, as book chapters are organized systematically for them to easily grasp the concepts. Additionally, the book contains exercises at the end of each chapter so students can evaluate their understanding of the topic. In general, this book is useful for anyone who is interested in the fundamentals of effort estimation in software systems or would like to pursue research in this domain.

I wish the very best to all the readers of this book.

Kirti Seth
Ashish Seth
Aprna Tripathi

MATLAB® is a registered trademark of The MathWorks, Inc. For product information, please contact:

The MathWorks, Inc.
3 Apple Hill Drive
Natick, MA 01760-2098 USA
Tel: 508 647 7000
Fax: 508-647-7001
E-mail: info@mathworks.com
Web: www.mathworks.com

Acknowledgments

This book is our voyage toward new knowledge and experiences. It has provided us the opportunity to present our abilities and motivation to conduct research and make recognized contributions to academia and industry. We owe our sincere thanks and appreciation to several people for their valuable contributions and suggestions to bring this work to press.

Firstly, we thank the **Almighty** for blessing us with strength and patience to overcome all the hurdles which we faced during the course of this work.

We owe our gratitude to our parents (Sh Vinod Tyagi and Smt Sushma Tyagi, Sh Harish Seth and Smt Krishna, Sh Ashok Kumar Tripathi and Smt Sushila Tripathi), and mother-in-law (Smt Asha Chaturvedi) for the blessings and encouragement they have provided during every endeavor of our lives. They are a continuous source of support and motivation.

We express our deep sense of gratitude and everlasting indebtedness to Dr. Arun Sharma, Department of Information Technology, Indira Gandhi Delhi Technical University whose advice, constant encouragement, invaluable suggestions, and expert guidance were extremely helpful in completing this book.

We are grateful to Dr. Lee Dong Won (First Vice Rector, Inha University in Tashkent, Uzbekistan) and Prof. A. K. Misra (former Professor, MNNIT Allahabad) for their continuing support and motivation which paved the way for the successful completion of this book. We extend special thanks to them for giving us constant encouragement and directions from time to time. We are thankful to all our colleagues at Inha University for providing valuable suggestions and advice which were very helpful in improving the quality of our work.

Kirti and Ashish Seth would like to dedicate this work to their lovely twin daughters Verda and Vindhya, for being "Good Luck" in their lives. Aprna Tripathi would like to dedicate her contribution in the book to her loving husband, Mr. Swamev Chaturvedi. Finally, we wish to express our thanks to all those people listed above as well those we did not mention, whose help and continuous inspiration made this work a little easier.

Kirti Seth
Ashish Seth
Aprna Tripathi

Authors

Kirti Seth, PhD, is a researcher and academician. She earned her PhD (Computer Science and Engineering) in the area of Component-Based Systems from the Department of Computer Sciences and Engineering, AKTU, Lucknow, India in 2016, MTech (Computer Science), from Banasthali Vidyapeeth, Banasthali, Rajasthan, India in 2009. She earned an MSc (CS) degree and has been involved in research and academics for the past fourteen years. She has published more than 40 research papers in reputed journals from renowned publishers such as ACM, Springer and Elsevier, and authored four books. Dr. Seth has participated in and organized seminars, conferences, workshops, expert lectures, and technical events to share knowledge among academicians and researchers and promoted opportunities for new researchers. She has provided training programs for students and faculties in various areas of computer science including Google's Techmaker event in 2018. She has given keynote addresses at many international conferences. Her current research interests include Component-Based Systems, Service Oriented Architecture, Bio-Inspired Optimizations, Neural Networks and Component-Based Systems, and Artificial Intelligence. She was bestowed the Young Scientist in Software Engineering award in 2017.

Ashish Seth, PhD, is a consultant, researcher, and teacher. He is a professor in the Department of Computer Science and Engineering, Inha University, Tashkent. He has more than 16 years of research and teaching experience. He has worked at several universities in India and abroad in various academic positions and responsibilities. He also serves as an editor, reviewer, and evaluator for reputed journals. He shares his knowledge through online forums; his educational channel on YouTube provides video lectures on varied areas of computer science; through an online forum he provides guidance to students on many subjects and helps them to discuss their problems. For his continuous dedication and contribution to enhancing the teaching and learning process he has been presented with Best Faculty award, Young Researcher award, and Most Promising Educationist award. His interests include reading and writing articles on emerging technologies.

Aprna Tripathi, PhD, began her academic career in 2002 teaching graduate courses in computer science at Engineering College Affiliated with UPTU (now known as AKTU) where she was an assistant professor. She earned her Master's from Banasthali University in 2009 and a doctorate from MNNIT Allahabad in 2015. Her major subjects of interest are software engineering, software testing, software architecture, big data, and data analytics. She has published more than 30 research papers in software engineering, with an emphasis on software metrics, effort estimation, and software engineering education.

Abbreviations

AD	Adaptability
AHP	Analytical Hierarchical Process
ANASSOC_CLS	Average Number of Associations per Class
ANCA_CLS	Average Number of Class Attributes per Class
ANFIS	Adaptive Neuro-Fuzzy Inference System
ANM_CLS	Average Number of Methods per Class
ANN	Artificial Neural Network
ANP_CLS	Average Number of Parameters per Class
ANREL_CLS	Average Number of Relationships per Class
CBD	Component-Based Development
CBS	Component-Based Systems
CBSD	Component-Based Software Development
CBSE	Component-Based Software Engineering
COM	Component Object Model
COM/DCOM	Component Object Model/Distributed COM
CORBA	Common Object Request Broker Architecture
COTS	Commercial off the Shelf
CP	Class Point
EE	Effort Estimation
EF	Environmental Factor
ESC	Extended Soft Computing
FC	Fuzzy Computing
FIS	Fuzzy Inference System
FL	Fuzzy Logic
FSVR	Fuzzy Support Vector Regression
FX	Flexibility
GA	Genetic Algorithm
IC	Interface Complexity
ISBSG	International Software Benchmarking Standards Group
LLE-SVM	Locally Linear Embedding and SVM
MF	Membership Function
MMRE	Mean Magnitude of Relative Error
MN	Maintainability
MRE	Magnitude of Relative Error
NOASSOC	Number of Associations
NOC	Number of Classes
NOCA	Number of Class Attributes
NOIR	Number of Inheritance Relationships
NOM	Number of Methods
NOP	Number of Parameters
NORR	Number of Realized Relationships

NOUR	Number of Use Relationships
OO	Object Oriented
OT	Object Orientation
OOD	Object-Oriented Design
OOP	Object-Oriented Programming
OMG	Object Management Group
PRED(L)	Prediction Level
RE	Relative Error
SDP	Software Defect Prediction
SVM	Support Vector Machine
TCF	Technical Complexity Factor
TDI	Total Degree of Influence
TSVM	Twin Support Vector Machine
UAW	Unadjusted Actor Weight
UCP	Use Case Point
UML	Unified Modeling Language
UUCP	Unadjusted Use Case Points
UUCW	Unadjusted Use Case Weight

1 An Introduction to Component-Based Software Systems

Improving business performance often requires improvement in the execution of product advancement, and this is the motivation that brings engineers and analysts closer to thinking about adopting the latest innovations and improvements. Previous systems evolved with the use of structured methodology, which was successful albeit only for straightforward applications. The Object Oriented (OO) approach came at that stage, which is based on encapsulation, inheritance, and polymorphism. Encapsulation combines the attributes and actions regulating the information in a single article. Inheritance engages a class, ascribes, and duties to be taken up by all subclasses and the items that they start from. Polymorphism allows different activities to have a similar name, reducing the estimation of the lines of code required to conduct a system. In the early 1990s, object-oriented programming became the main component of decisions made by some product manufacturers, a number of data frameworks, and experts in building. In spite of this, the core behavior of software engineers is still unchanged except for a few OO approach preferences: composing line-by-line code. In case of difficulties, it is not easy to implement object-bearing to complex applications. In addition, OO dialects reinforce class-level coverage of results, but not beyond that. Certain fundamental issues in the OO approach are integrity and privacy.

1.1 COMPONENT-BASED DEVELOPMENT

This is an approach that recognizes the fact that many data-based systems include similar or even indistinguishable items that are generated over and over without any planning. Progress is increasingly costly from the start and can require a lot of effort to finish. Because of postponements in the improvement process, simple applications with severe time constraints can liberate the market. This has facilitated the development of another approach, called component-based development (CBD), which uses the reusability concept to improve the application. The use of regular components enables higher profitability. The two core values which accompany CBD can be best depicted as:

(1) Re-use but do not reinvent.
(2) Instead of coding line-by-line, assemble pre-built components.

Here, sections are seen as groups of schedules that are developed based on well-characterized criteria, with the aim that these families work together as blocks of construction.

1.1.1 COMPONENT

Component-based software development (CBSD) is a technique in which software is worked by combining the components with individual parts from freely accessible items. A few proposals emphasize that components are handily informed packets about useful behavior, while others ensure that components are real, deployable programming units that are implemented within a system.

A few concepts for the components have been suggested by researchers. The majority of those are discussed in Table 1.1.

In short, a component with a well-defined interface is a reusable, independent piece of software that is self-supporting to any task. The notable point to remember when constructing a component is the concept of reusability, whether or not a combination will recognize what the segment's potential needs will be. Components may be placed on any device core, depending on the preconditions for operation and regardless of the device's particular structure. In order to make the section usable for longer, effort must be made to ensure utility of the component beyond the needs of the current application (Gill and Grover, 2003).

TABLE 1.1
Different Definitions of Components

Researcher	Definition
Sparling, 2000	A component is a non-partisan language, freely executed bundle of software services, transmitted in an embodied and replaceable compartment, obtained through at least one distributed interface. While a variable may have the ability to alter a database, it should not be dependent on holding the stated information. A category is not restricted to coordinating, nor is it app friendly.
Szyperski, 2002	A software component is a synthesis unit with authoritatively defined configuration and explicit setting conditions. A software component can be freely distributed and third parties exposed to synthesis.
Microsoft Corp	A software component is an integrating, distribution, or conveyance device that provides styles of assistance within a data term adherence or limit of encapsulation.
D' Souza and Wills, 1999	A part of a commodity is a kind package of software execution which can be independently generated and imparted. It has expressed and well-shown interfaces that it foresees for the services this backs up and for the different services.

1.1.2 GENERAL COMPONENT PROPERTIES

Modularity should be used as the basis for system design to develop good components. The criteria for determining a framework's modularity potential are defined in Table 1.2.

In a company with its own lifecycle, to obtain this option, a component must have its own:

(1) Set of requirements.
(2) Set of documents.
(3) Test cases.
(4) History (board arrangement), which distinguishes between the software of the component and the perspectives referred to above.

The life pattern of a section would naturally be conceivable only along these lines. Through preserving an individual design background for each component, an application may "decide" whenever it becomes available, regardless of whether to move up to another component type. Perhaps the aim of understanding the "part versus

---

TABLE 1.2
Modularity Criteria of a Component

Criteria	Definition
Decomposability	A component-oriented approach to change will assist the product designer in characterizing a framework made up of less complex subsystems with a clear structure in between. Recognized subsystems (segments/objects/modules/...) may be sufficiently free to allow for a similar property for decomposability on each of them.
Composability	The component should be done in order to improve the reusability in different settings. Depending on their interface, each component may simply be freely joined with others.
Understandability	The framework element structure must help the comprehensibility of each of the components. In order to see each part, one might just need to take a quick look at the component itself and not in the frame into which it is incorporated.
Continuity	Continuity means that an obstacle to a neighborhood has a consequence nearby. Continuity is a major worry of programming maintenance as it could impact the cost of programming development. The law of movement has to be an important goal when developing new components.
Protection	In order to ensure the stability of the entire system in which components are stored, each component must ensure that not all of the structure causes mistakes or some other irregular conditions.
Self-containment	The concept of the component has different meanings. Altogether it has to be free of any requirement for a section to be reusable without restrictions. This means that each part should be viewed as a small, independent part.

object" approach is that one can concentrate on business objectives around area development when considering a section. Usually, the term "component" is applied to assign the company somehow to a self-servicing bit of code with services of key significance.

1.1.3 COMPONENTS AS OBJECTS AND FRAMEWORKS

The development of systems based on components starts with a range of pre-existing parts. The components are embedded with some restrictive code in the frame that holds them together. This code is called the "glue code". Object orientation (OT) has a similar methodology; objects are interchangeable components that can be combined into programs. OT is not appropriate for component-based software engineering (CBSE). OT does not pass on the use's relationship, which is important for CBSE. OT express "has an" and "is a" association. Components pass on the condition in which they will work by demonstrating what framework assets the segment needs to work appropriately. OT for the most part does not back up this kind of idea. A component need not be a thing; it might just be a capacity or an executable program which is not treated as an object. In any case, regardless of whether items (or classes) cannot be seen as a feature of the plan, they are reusable segments and OT inventiveness can be utilized successfully to upgrade parts. Object-oriented design (OOD) structures are increasingly favored for upgrading programming contrasted with publications. Structures are software components containing objects related with, and in a described setting by, shared relationships. A Framework portrays a utility that is at a more significant level of abstraction. The motivation behind why Frameworks are favored over articles is that objects can, as a rule, have more than one occupation in more than one condition; OOD systems can do that, and OOD structures can get this; however, existing OOD techniques cannot. The last use classes or items as the essential unit of structure or reuse, and depends on the customary perspective on an article, as appears in Figure 1.1 which views an article as a closed substance with one fixed job.

Systems generate protests in various frameworks that presume different jobs. For example, in Catalysis this is delineated in Figure 1.2.

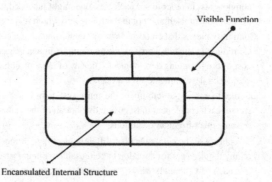

Visible Function

Encapsulated Internal Structure

FIGURE 1.1   Objects with one fixed role.

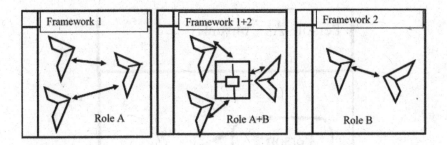

FIGURE 1.2 Objects with multiple roles in different frameworks.

The accompanying model represents the system idea. Think about the structure for workers as delineated in Figure 1.3 in which an individual assumes the job of a worker of an organization.

A person as a representative has a property pocket which speaks to the cash measure he has, and two activities receive pay and work. Right now, consider an alternate point of view of an individual, for example an individual assumes the activity of a consumer, as appears in the Person As Consumer structure in Figure 1.4. Equally, the individual right now has the property pocket, but he has the purchase of the activity (instead of paying and working for the activities). For Person As Employee and Person As Consumer, we may form the structures to acquire an individual with the two jobs combined. An individual currently has all the activities of both jobs, including receiving pay, working and buying, task and purchase, and the characteristic pocket in the two jobs. Figure 1.5 illustrates the composition.

Frameworks' fundamental qualities are the plausibility of manufacturing Frameworks from halfway items, where an article assumes explicit work. Equally,

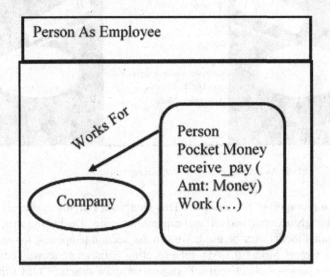

FIGURE 1.3 Person As Employee framework.

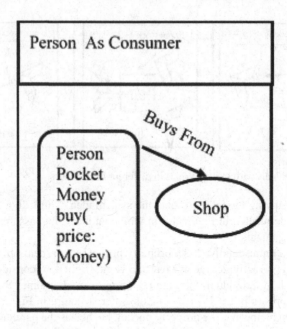

FIGURE 1.4   Person As Consumer framework.

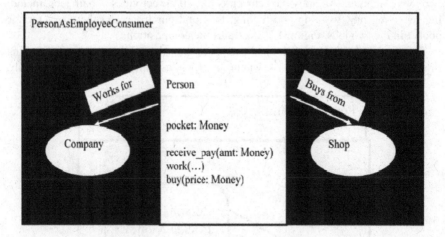

FIGURE 1.5   Person As Employee Consumer framework.

systems can incorporate Frameworks. This idea makes producing larger segments with increasingly accurate and rational capabilities easier. During the structure stage this conceptual model may be used. We can use section templates, for example the Component Object Model (COM), for executing parts as systems. The following model shows how to collect "partial" papers within a system. COM suits various tasks since it can use different interfaces for each job.

Second, buyers and representative jobs are updated to support being accumulated into a single item. Additionally, the purchaser object needs a reference to the individual article in order to have the option to take a shot at the pocket variable. The reference Person shall be identified when the buyer is accrued in the individual document. The Employee job is done along these lines. We may reuse the various parts that we have created using the conglomeration instrument in COM. Figure 1.4 shows how Person aggregates the two COM things previously described, Customer and Employee, which assume different duties of a specific person. Systems are rendered by adding jobs to a document at runtime. The entire Person could now be treated as a single component. Interfaces become the standard interface for each acquired object.

There are a few constraints to the COM implementing the program concept. The COM model characterizes systems as the total of the finished items generated during runtime, while a general framework model allows us to use deficient items (at runtime) or classes (at manufacturing time).

1.2 COMPONENT-BASED SOFTWARE ENGINEERING

Code is reused as objects in object-oriented programming (OOP), and a couple of devices, for example, inheritance and polymorphism, let the designers reuse these in various ways. Thus, the standard is like component-based software engineering (CBSE), however here the emphasis is on reusing the entire programming segment, not simply the objects.

CBSE is a way of thinking that plans to make and sort out structures utilizing a pre-portrayed assortment of programming parts explicitly intended for reuse. This moves the complement from programming to confining the programming. As Clements (1996) has shown, CBSE exemplifies "the 'purchase, don't assemble' reasoning".

1.3 ADVANTAGES OF COMPONENT-BASED SOFTWARE ENGINEERING

Following are the main advantages of CBSE:

- **Flexibility.** Runtime parts can operate autonomously and are considerably less subject to their condition (equipment, system programming, different applications, or components) when designed accordingly. Therefore, component-based frameworks are far more flexible and extensible than custom-structured and assembled frameworks. As a rule, components are not altered, but are supplanted. That adaptability is significant in two regions:
- **Hardware and system software.** Component-based frameworks are less vulnerable to changes in the system (for example, the working framework) than traditional frameworks. This results in an increasingly rapid movement from one working system to the next or from one DBMS to the next. Furthermore, an interesting finding is the possibility of a frame in a currently heterogeneous state.

- **Functionality.** Component-based frameworks are far more flexible and extensible at a useful level than normal frameworks, on the grounds that a large portion of the new functionality can be reused or derived from pre-existing parts.
- **Reusability.** On a basic level, CBD empowers the enhancement of parts that completely implement a particular agreement or business objective. Those components can be used everywhere. Usefulness, whether specialized or structured, should be generated and revised only once, rather than several times, as is generally the case at present. It will be clear that from the viewpoint of practicality, rhythm, and competitiveness, this is something to be grateful for. Clearly, this reuse can be within an entity, but more than just a few organizations.
- **Maintainability.** Ideally, a bit of utility is executed only once in a component-based system. Such findings are evident in easier management, which leads to lower costs and a longer life for these systems. Truth be told, it will turn out that the distinction between upkeep and development is ambiguous and will completely vanish after a while. A high proportion of new applications must contain existing components. Building a system will look more like coming together than actually putting together. Additionally, the big, strong frameworks of which we are probably aware, will vanish, allowing the differences between the frameworks to become obscured.
- **More rapid development/higher productivity of the developers.** In principle, for reuse purposes, among others, CBD will bring about an increasingly rapid improvement in the frameworks. The higher performance will be worked out over the medium to long term. Be that as it may, the results of reuse will be smaller in the short run than the cost of providing another approach to enhance the system. In addition, reusable segments are not sufficiently accessible at present.
- **Distribution.** CBD allows the structuring of dispersion systems (on a LAN, or on the web, etc.). In fact, this is one of the primary aims of the present promotion. It is also, naturally, an extremely advantageous situation.

Several product category models are available on the market, like Microsoft's Component Object Model (COM), DCOM, Enterprise Java Beans (EJB), J2EE, and Object Management Group's (OMG) Common Object Request Broker Architecture (CORBA) information, .NET Framework, Sun's Java Beans, etc.

In CBSD, as in different types of uses or systems, a significant number of the methods, tools, and principles of software design can be used in a similar way. Be that as it may, one distinction exists, CBS encompasses both component development and system development. Additionally, the needs and the path to progression are somewhat distinct. Components are worked for use and reuse in various applications, some of which are not yet available. A part should be highly indicated, concise, wide enough, easy to adjust, easy to convey and submit, and easy to supplant. The component interface must be as simple as can be reasonably expected and carefully separated from its implementation (both logically and physically). Showcasing variables require substantial work, since the cost of development must be recovered from future income.

Improvement with components centers around the identification of reusable elements and the relationships between them, starting from the preconditions of the framework.

Component-based development (CBD) is based on having a brilliant future and a large scope to explore. Current research in theory contemplates and examines a few consistency highlights of structures focused on sections and pieces, and provides metrics for multifaceted value parts of design, reusability, and viability. To organize these measurements, it uses Artificial Neural Network, Analytical Hierarchical Method, and Fuzzy Logic. In addition, it suggests a consistency model for CBS and tests it from live undertaking on a concrete investigation.

1.4 CONVENTIONAL SOFTWARE REUSE AND CBSE

Despite the fact that advances in object situations have facilitated software development reuse, there is a big gap between the whole frameworks and classes. For quite a long time many fascinating thoughts have risen in object-situated programming reuse to fill the gap. We combine computer engineering (Shaw and Garlan, 1996; Gamma, Helm, Johnson, and Vlissides, 1995), structure designs, (Fayad and Schmidt, 1997), and systems.

1.4.1 CBSE APPROACH

CBSE adopts various strategies from ordinary programming reuse.

(1) **Plug and Play:** Component should be able to connect and play with other components and/or frameworks, so that components can be constructed without compilation at runtime.
(2) **Interface-centric:** Component can distinguish the interface from the implementation and cover information about implementation, so that they can be constructed without understanding how to implement them.
(3) **Architecture-centric:** Components are built to interoperate with other components and/or frameworks on a predefined architecture.
(4) **Standardization:** Component design should be standardized so that several vendors can produce them and reuse them widely.
(5) **Distribution through the market:** Components can be acquired and improved though a competitive market and provide incentives to the vendors (Figure 1.6).

Table 1.3 summarizes major characteristics of conventional software development and component-based software development, which are briefly discussed in the following section

1.5 ARCHITECTURE

Many component-based systems, for example Microsoft Foundation Class (MFC) and CORBA, expect encapsulated software plans. They are handled as systems.

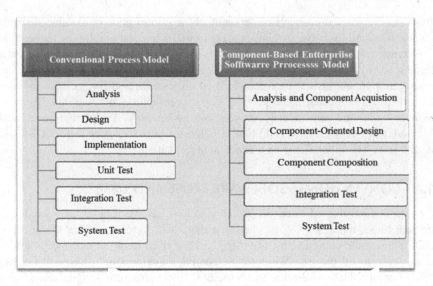

FIGURE 1.6   Conventional software reuse and CBSE.

TABLE 1.3
Characteristics of Conventional Software Development and Component-Based Software Development

Characteristics	Conventional	CBSE
Architecture	Modular	Monolithic
Components	Interface and Black-Box	Implementation and White-Box
Process	Evolutional and Concurrent	Big-Bang and Waterfall
Methodology	Composition	Build from Scratch
Organization	Specialized: Component Vendor, Broker, and Integrator	Monolithic

Structures are a practical guide when developing basic programming. To be effective, the structure can be stepped up differently from the territory which makes unambiguous space models fuse the Eagle Adventure (Ande98) by Andersen Consulting and the San Francisco Adventure (IBM98) by IBM (Lazar, 1998). Through uniform and projected build-up of programming, the CBSE can avoid adhoc design.

1.6   PROBLEMS AND PITFALLS OF CBD

This is certainly a rather interesting list of attractions. Here the question arises: Will components contain any downsides or challenges? Yes, indeed. To begin with, an appropriate situation is critical on the off-chance you need to make the best use of pieces. Clearly, rapidly producing your own conditions over the working system is conceivable in theory. But in practice, it is not possible. In fact, reuse is one of the most motivating aspects for using components, meaning it should be conceivable to reuse a component created for another purpose, probably in another environment. It implies standardization, and in fact this is the biggest obstacle to CBD's performance right now. Benchmarks are expected regarding the middleware the components will operate in. Middleware can mean a lot of things, but what is inferred here is a layer of communication that empowers parts to send more substantive-level messages to different segments, if appropriate in a system. Obviously, what has a place in the middleware, and so forth, is open to discussion. For example, should the dealing with trade be a piece of it? What about commitment? There are currently three contending frameworks on this field: CORBA from the Object Management Group, Microsoft's COM/DCOM and Java's Remote Method Invocation. In addition, these three norms are not exactly identical to each other. CORBA is a standard only, while COM/DCOM is not only a standard but also a function. CORBA is actually an independent language, COM/DCOM and RMI, as far as anyone knows, are a part of the Java standard.

1.7 FIVE PROBLEMS OF EFFORT ESTIMATION

In CBS systems, the problems with respect to effort estimation can be explained through five questions:

(1) Why are legacy techniques for cost estimation not valid for CBS?
   It is always possible to underestimate or overestimate efforts for components, as these components are pure design artifacts.

   Effort estimation of CBS depends upon many criteria. The two most important criteria are as follows:
   • Interaction among components.
   • Complexity of these interactions.

These interactions may cause interdependencies among components. CBS complexity is directly proportional to these dependencies.
   Another factor which might be considered for complexity is the type of integration. Hence, it is not easy to estimate efforts for CBS.
   The complexity of CBS is much higher as compared to legacy systems, because of these reasons effort-estimating techniques, available for traditional systems do not work for CBS.

(2) What are the consequences for not evaluating efforts accurately?
   Some important consequences for incorrect estimation of efforts are:
   • **Cost increment:** If efforts are not estimated correctly then either costs may be excessive, or in case of underestimation, costs may be reduced.
   • **Delivery delay:** The second result may be a delay in delivery. This may be possible either because of delivery or resource constraints.
   • **Withdrawal:** It may be the chance that customer withdraws projects.

Social consequences are goodwill and customer dissatisfaction.

(3) Who are most suitable team members to guide the budget?
Members of a team according to their role are known as project owners, developers, architects, profit managers, etc.

Developers are the individuals who are responsible for coding, they may give valuable suggestions to the owners.

The project manager is responsible for evaluating the method for assessment given by the process owner for evaluation of project.

(4) What is the correct time for reevaluating the expectations?
Down the line, project scope and planning may change; things may not go as expected. When it is clear, there is a need to reevaluate the expectations of the technology, the people, the condition of the platform, and the risks to the environment or specific activities.

(5) What are the areas for effectively assessing efforts?
The following areas are suggested:
• Scheduling and staffing.
• Managing cost increment during the project.
• Completion of project in time within a previously defined budget.

EXERCISE

Question 1. In which scenarios is component-based development better than object-oriented development?

Question 2. Although CBSE claims a high degree of reusability in comparison to OO Software Engineering, it invites another level of complexity during the maintenance phase. Suggest some solutions to reduce such complexity.

Question 3. What are the essential contrasts between objects and components?

Question 4. Using an example of a component that implements an abstract data type, show why it is necessary to extend and adapt components for reuse.

Question 5. Design a reusable component that implements the search feature for online shopping. This is not a simple keyword search of web pages. You have to be able to search the item by different categories, as specified by the user.

Question 6. Why it is important that components should separate the interface from the implementation and hide the implementation details?

Question 7. What are the challenges in component distribution through the market?

REFERENCES

Ande98. Retrieved from http://www.ac.com/aboutus/tech/eagle.

Clements, P. C., & Northrop, L. M. (1996). *Software architecture: An executive overview (CMU/SEI-96-TR-2003).* Pillsburgh, PA: Software Engineering Institute, Carnegie Mellon University.

D'Souza, D. F., & Wills, A. C. (1999). *Objects, components, and frameworks with UML – The catalysis approach.* Reading, MA: Addison-Wesley.

Fayad, M. E., & Schmidt, D. C. (Eds.). (1997). Object oriented application frameworks. *CACM, 4 32–38.*

Gamma, E., Helm, R., Johnson, R., & Vlissides, J. M. (1995). *Design patterns: Elements of reusable object-oriented software.* Reading, MA: Addison-Wesley.

Gill, N. S., & Grover, P. S. (2003). Component –Based measurement: Few useful guidelines. *ACM SIGSOFT Software Engineering Notes, 28*(6), 1–4.

Lazar, B. (1998, February). IBM's San Francisco project, software development, *6*(2), 40–46.

Rubin, B. S., Christ, A. R., & Bohrer, K. A. (1998). Java and the IBM San Francisco project. *IBM Systems Journal, 37*(3), 365–371. 10.1147/sj.373.0365.

Shaw, M., & Garlan, D. (1996). *Software architecture: Perspectives on an emerging discipline.* Englewood Cliffs, NJ: Prentice-Hall.

Sparling, Michael (2000). Lessons learned through six years of component-based development. *Communications of the ACM, 43*(10), 47–53.

Szyperski, Clemens (2002). *Component software: Beyond object-oriented programming* (2nd ed.). Addison, TX: Wesley Longman Publishing Co., Inc.

2 Effort Estimation Techniques for Legacy Systems

2.1 INTRODUCTION

In the case of project management, effort estimation in software development is an emerging trend. Effort estimation will come under the analyzed variables. In cases of considering prediction and estimation there are some unresolved issues. By adopting higher reliability, the process of estimation of effort is not yet possible. The process of estimating effort is still a tedious process for project managers.

For processes like project planning, project budgeting, and project bidding the estimation of effort in software projects is considered fundamental. This effort estimation is taken seriously in software companies because inaccurate planning and budgeting can lead to serious consequences. If negative plans and budgets are made the software industry will face a loss of business opportunities. Effective cost estimations are necessary for software development. Some of the work related to this determination of estimation effort is reviewed as follows.

2.2 THE IMPORTANCE OF PRECISE EFFORT ESTIMATION TERMINOLOGY

The development of accurate and standardized techniques will not effectively solve the issues related to effort estimation. The necessary importance must be placed on the technique for solving problems easily and effectively. At the same time, a high degree of precision must be applied. Examples of the most likely problems are illustrated in Table 2.1.

A comparison analysis was conducted for the performance of effort estimation for the developed technique and for the techniques which are surveyed. It is not possible to obtain a better result because there will be variations in sizes and types.

2.3 TRADITIONAL TECHNIQUES OF EFFORT ESTIMATION

Traditional software models are directed towards large monolithic software development projects. In a component-based software development environment, effort estimation models can benefit from a fine-grained view of parameters that influence effort. Software development can be viewed through a structured programming paradigm. The projection of software costs is made at a general system level for these models.

TABLE 2.1

List of Problems Occurring Due to Lack of Precision

Sl. No.	Problem	Example
1.	A mixture of procedures for various reasons.	A combination of procedures with attention to authenticity (estimation of in all probability exertion), an emphasis on productive improvement work (estimation of arranged exertion), an attention on shirking of spending (choices on planned exertion), and an emphasis on winning an offer (estimation of cost to-win). The absence of partition between these procedures has been found to decrease the authenticity of estimation in all probability exertions. .
2.	Estimation error comparison for various software projects when they are not actually comparable.	One of the software projects may have made an estimation error, estimation with respect to the distinction between arranged exertion and genuine exertion, while another task may have made theirs with respect to the contrast between all likelihoods of exertion and real exertion.

2.3.1 RULE OF THUMB

This method is used for estimating effort based on practice and previous experience. It is an easily learned and easily applied procedure. This method is not necessarily an accurate or a reliable method as it approximates effort based on past experience without any formal basis. The following rules of thumb, as given in Fairley (1992), can be utilized:

(1) Expected productivity in terms of lines of code per programmer month.
(2) Quality in terms of errors per thousand lines of code.
(3) Percentage distribution of effort and schedule for various activities like design, coding, analysis, etc.

Although many rules of thumb are in use today because they were simple to implement, they can never be replaced by formal estimation methods. Another set of rules of thumb has been developed for forecasting software growth rates during maintenance, optimal enhancement size, and annual software enhancements (Jones, 1996; Smith, 1999).

2.3.2 ESTIMATION BY ANALOGY

In cost estimation methods, a comparison evaluation is made between completed projects and new projects within similar areas of application. From the database for creating the estimate the same sorts of project were chosen in the technique called Estimation of Analogy. Through this analogy approach the cost for the new project was estimated with the assistance of the completed project (Bohem, 1981).

The following are the issues of concern while using the estimation by analogy approach:

(1) Right function selection to determine the degree of similarity between target and new project.
(2) Relevant projects selection for estimation.
(3) Deciding the number of similar projects to be used for estimation.

According to Construx Software Inc., it was found that estimation carried out on the basis of historical data obtained from the organization was more precise when compared to the estimation based on educated guesswork (Construx Software) and rules of thumb. Many organizations do not have sufficient data to use this method of estimation. To overcome this problem, the International Software Benchmarking Standards Group (ISBSG) developed certain metrics for software projects that are found to enhance their benchmarking, productivity, risk analysis, and project estimation most effectively. For performing the evaluation on data repositories several software tools and publications were produced by ISBSG. Based upon metric types certain data will be provided through ISBSG. With the assistance of this data the information regarding past projects in the organization will be gathered and utilized in future estimations. The common entities for which organizations build metrics are

(1) Products that are estimated using product metrics.
(2) Processes that are estimated using process metrics.
(3) Qualities that are predicted using quality metrics.

2.3.3 FUNCTION POINT METHODS AND THEIR LIMITATIONS

In 1979 Albrecht developed a conception for function points. This function points come under the metric called application size (Kitchenham, 1997). Through this function point any effort present in the application can be detected.

Five main kinds of fundamental elements are present in function points, which include External Interface File (EIF), External Inquiry (EI), Internal Logical File (ILF), External Output (EO), and External Input (EI). For every element, the complexity is classed as high, average, or low.

The Complexity Adjustment Factors (CAF) and Unadjusted Function Points (UFP) are multiplied to obtain the function point. The equation for this function point is represented as follows.

$$FP = UFP \times CAF$$

The summation of factor of complexities (*Wij*) multiplied with number of classified elements (*Iij*) within the application is referred to as *UFP*. The equation for this UFP is represented as follows:

$$UFP = \sum_{i=1}^{5} \sum_{j=1}^{3} Wij \times Iij$$

TABLE 2.2

Weights of I_{ij}

FUNCTION UNITS	AVG	LOW	HIGH
Internal Logical Files	10	7	15
External Output	5	4	7
External Interface File	7	5	10
External Input	4	3	6
External Inquires	4	3	6

Where W_{ij} is the weight of I_{ij} and Table 2.2 represents the values of W_{ij}.

For the entire application the process complexity and environment was evaluated through CAF. The features of 14 basic systems have some influence and this is rated on a scale between 0 and 5 (5 = Essential, 4 = Significant, 3 = Average, 2 = Moderate, and 1 = Incidental). This rating was given based on its effect in the project.

Then CAF is computed using following formula:

$$CAF = 0.65 + .01 \times \sum f_i \text{ where I = 1 to 14}$$

Example 1:

A project is considered along with its subsequent functional units:

- Number of external outputs = 4.
- Number of external inputs = 5.
- Number of external interfaces = 2.
- Number of external enquiries = 3.
- Number of Internal files = 2.

It is assumed that all the weighting factors of elements are simple additional complexity adjustment factors and were found to be average. For the given project a function point is evaluated.

SOLUTION

Function Point Elements	Number of Elements (I_{ij})	Weights of Element (W_{ij})	$W_{ij} \times I_{ij}$
Number of external inputs	5	3	15
Number of external outputs	4	4	16
Number of external enquiries	3	3	9
Number of external interfaces	2	5	10
Number of Internal files	2	7	14
UFP =			**64**

$$CAF = 0.65 + .01 \times 14 \times 3 \left(\text{since all the 14 factors are average}\right)$$

$$= 0.65 + .01 \times 42$$

$$= 0.65 + 0.42$$

$$= 1.07$$

$$FP = UPF \times CAF$$

$$= 64 \times 1.07$$

$$= 68.48$$

The drawbacks and complications that were found while utilizing the function point method is explained by Kitchenham in his article. In the previous section it was explained that the complexity of the element was classified as high, average, and low. This implies that the measurement was using an ordinal scale and the use of absolute scale counts was declined. Basically, the measurement that is obtained from an ordinal scale is impossible to add together because the elements such as high, average, and low are separate matters. In case of utilizing similar factors, adding cannot be achieved. In an ordinal scale the measurement of size can be achieved. It can say that the size of one particular system is bigger when compared to another but the value for the size cannot be found through this function point approach. The number of elements to be considered will be decided through examining the Symons Mark function point construction. It was found that for equating the entity, output, and input no standard conversion factors exist (Kitchenham, 1997).

The correlation between the elements in the function point must be found. This is considered a significant criterion. The function point is viewed to be inconsistent, which is another limitation.

2.4 EFFORT ESTIMATION FOR OBJECT-ORIENTED SYSTEMS

In the process of software development, the technique of Object-Oriented Systems has emerged in recent years.

Consequently, many researchers have proposed several metrics suitable for measuring the size and the complexity of object-oriented software, assessing aid, or introducing new metrics that capture exclusive concepts of the object-oriented paradigm, such as inheritance cohesion and coupling. Traditional metrics, such as Function Points, are unsatisfactory for predicting software size because they are based on the procedural paradigm that separates data and functions, while the object-oriented combines them.

Three dimensions were established in object-oriented software. The metrics of object-oriented systems must follow these dimensions:

- Percentage of reuse through Inheritance.
- Amount of communication among objects.
- Functionality (behavior of objects).

Various object-oriented approaches are as follows:

2.4.1   UML-Based Approach

The second approach to effort estimation of CBSD (Component-Based System Development)is Unified Modeling Language (UML) based on the object-oriented approach. In this approach the main characteristic is the use of case points.

Through the conception of object-oriented systems the software system can be described using visualization. UML is a technique developed based on this object-oriented approach. This is referred to as the de facto standard. The above-mentioned UML does not come under an architecture description language, but through UML diagrams the necessary data can be collected and used in size and complexity measurement of the software system. This language-based measurement can be very effective in the upstream level of software designing.

Use case points:

The use case point method was developed for performing effort estimation in the initial stage. Use case point (UCP) measurement can be done through use case modeling. The functional scope required for the development of a software system can be defined with the help of the use case model. This model is not related to the function point method and it depends purely on use case point. The advantages of this method for estimating effort in the initial stage has been explained in various articles.

Counting use case points: Within the flow of events, the number of transactions and actors involved will be counted along with a certain weight for measuring the UCP. When it is conducted among the target systems and actors, it is called a transaction. Within the use case model, the represented actor is defined as a simple, complex, or average. By means of a simple actor another system will be denoted as API. The interaction carried out with another system by means of TCP/IP protocol or the interaction carried out by the person utilizing a text-based interface will be referred to as an average actor. The interaction of the person will be attained through GUI interface in the case of complex actors.

The steps followed in the UCP method for estimating effort are represented as follows:

Step 1: (Counting the actor's weight).

In the target software system, the number of actor types involved will be counted and this number of actor types will be the product with a weighting factor. The actor weighing factor is represented in Table 2.3. By summing these values the actor weight is measured (Somerville, 2001; Martin Shepperd et al. 1996).

Step 2: (Counting use case weight).

The categorization of every use case will be complex, average, or simple. The decision will be taken based on the number of transactions made in the use case. The alternative path is also included in this transaction. The definition of a transaction is that it is referred to as set of activities, it will be carried out as a whole or not. The

TABLE 2.3
Factors Related to Actor Weighting

Factor	Description	Type
3	Graphical interface	Complex
2	Interactive or protocol driven interface	Average
1	Program interface	Simple

use case which is referred to as extended or included is not taken into account. It was found that seven transactions will be seen as a case of complex use case, four to seven transactions will be seen as a case of average use case, and three or fewer transactions will be seen as a case of simple use case.

In the target software systems, the number of use case types involved will be counted and this number of actor types will be the product with a weighting factor. In Table 2.4 the actor weighing factor is represented. By summing these values the use case weight is measured

Step 3: (Calculating UUCP).

The total for Unadjusted Use Case Weight(UUCW) and total weight for actors, Unadjusted Actor Weight (UAW), will be summed to obtain Unadjusted Use Case Points (UUCP).

$$i.e, UUCP = UUAW \times UAW$$

Step 4: (Weighting of environmental and technical factors).

On the basis of the value allocated for the number of environmental and technical factors, the obtained UUCP will be altered. The values for the number of environmental and technical factors is represented in Tables 2.5 and 2.6. By calculating the effect of the Ach factor in the application a value is allotted in the range of zero to five. It implies that if the rating is found to be five then the factor is necessary for the application, in cases of a rating of zero then it is determined to be irrelevant.

TABLE 2.4
Transaction-Based Weighting Factor

Factor	Description	Type
15	More than 7 transactions	Complex
10	4 to 7 transaction	Average
5	3 or fewer transaction	Simple

TABLE 2.5
Technical Factors Given for Weight and System

Weight	Description	Factor Number
1	Special user training facilities are required	T13
1	Provides direct access for third parties	T12
1	Includes special security features	T11
1	Concurrent	T10
1	Easy to change	T9
2	Portable	T8
0.5	Easy to use	T7
0.5	Easy to install	T6
1	Code must be reusable	T5
1	Complex internal processing	T4
1	End-user efficiency	T3
1	Response or throughput performance objectives	T2
2	Distributed System	T1

The product of weight and the value of every factor represented in Table 2.5 (T1–T3) will be made first to obtain the technical factor (TCF). The obtained value will be summed to determine the technical factor. The equation used for the technical factor is represented below:

$$TCF = 0.6 + (0.01 \times TFactor)$$

The product of weight and value of every factor represented in Table 2.6 (F1–F8) will be made first, to obtain the environmental factor (EF). The obtained value will be summed to determine the E Factor. The equation used for the environmental factor is represented below:

$$EF = 1.4 + (-0.03 \times EFactor)$$

Step 5: (Evaluating UCP).

Equation for finding the UCP is represented as follows:

$$UCF = EF \times TCF \times UUCP$$

Step 6: (Effort estimation).

Finally, the obtained UCP and specific values, which are represented in man-hours, will be multiplied to measure the estimation effort.

TABLE 2.6

Environmental Factors Represented for Weight and Team

Weight	Description	Factor
−1	Difficult programming language	F8
−1	Part time workers	F7
2	Stable requirements	F6
1	Motivation	F5
0.5	Lead analyst capability	F4
1	Object-oriented experience	F3
0.5	Application experience	F2
1.5	Familiar with the national unified process	F1

Example:

The utilization of a Restaurant Management System (RMS) will be made to perform the operation of UCP measurement. The generation of a Use Case Diagram for a particular software system is represented in Figure 2.1.

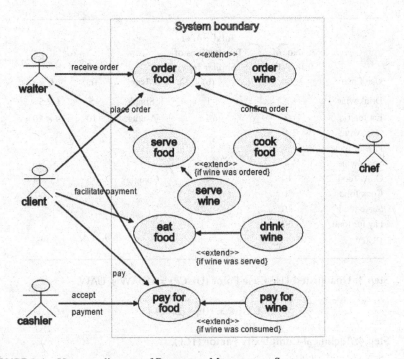

FIGURE 2.1   Use case diagram of Restaurant Management System.

Make the following assumptions for the components utilized in the use case diagram shown in the figure: one actor is of type simple as chef, one actor is of type average as chef and other two are "complex" as client and cashier. Order food, cook food, serve food and pay for food have ten transactions each – eat food has seven transactions – order wine, serve wine and pay for wine have three transactions each and drink wine has two transactions. Make the following assumptions for the system as a whole: Assigned Value for all the Technical Factors is two and for all eight Environmental Factors is three. Calculate the UUCP for utilizing the RMS. Also, calculate the total effort for developing the RMS if eighteen man-hours per use case point is utilized.

Step 1: Unadjusted Actor Weight (UAW).

Evaluation performed for finding UAW is represented below:

$$UAW = (\text{Total No. of complex Actors} \times 3) + (\text{Total No. of average Actors} \times 2)$$
$$+ \text{Total No. of Simple Actors} \times 1)$$

$$\text{For RMS, UAW} = (2 \times 3) + (1 \times 2) + (1 \times 1) = 6 + 2 + 1 = 9$$

Step 2: Unadjusted Use Case Weight (UUCW).

Use Case	No. of Transactions (a)	Total No. of Transactions of Similar Type (b)	Type	Factor (c)	b × c
Drink wine	2	1	Simple	5	$1 \times 5 = 5$
Eat food	7	4	Average	10	$4 \times 10 =$
Oder wine	4				40
Pay for wine	4				
Serve wine	4				
Order food	10	4	Complex	15	$15 \times 4 =$
Cook food	10				60
Serve food	10				
Pay for food	10				
UUCW					105

Step 3: Unadjusted Use Case Point (UUCP) = UUAW + UAW.

$$UUCP = 105 + 9 = 114$$

Step 4: Technical Complexity Factor (TCF).

Given that the assigned value to each technical factor is 2, therefore:

T. Factor	No. of TF of Similar Type (a)	Weight (b)	Assigned Value (c)	a × b × c
T1 and T8	2	2.0	2	$2 \times 2 \times 2.0 = 8.0$
T2–T5, T9–T13	9	1.0	2	$9 \times 2 \times 1.0 = 18.0$
T6 and T7	2	0.5	2	$2 \times 2 \times 0.5 = 2.0$
TF				28.0

$$TCF = 0.6 + (TF/100)$$

$$\text{For RMS, } TCF = 0.6 + (28/100) = 0.88$$

ENVIRONMENTAL COMPLEXITY FACTOR (ECF)

Given that the assigned value for each environmental complexity factor is 3:

E. Factor	No. of TF of similar Type (a)	Weight (b)	Assigned Value (c)	a × b × c
F1	1	1.5	3	$3 \times 1 \times 1.5 = 4.5$
F2 and F4	2	0.5	3	$3 \times 2 \times 0.5 = 3.0$
F3 and F5	2	1	3	$3 \times 2 \times 1 = 6$
F6	1	2	3	$3 \times 1 \times 2 = 6$
F7 and F8	2	–1	3	$3 \times 2 \times -1 = -6$
EF				13.5

$$ECF = 1.4 + (-0.03 \times 13.5)$$

$$\text{For RMS, } ECF = 14 - .405 = 0.995$$

Step 5: Use Case Points (UCP).

The evaluation of Use Case Points (UCP) can be performed with the following formula:

$$UCP = (UUCP) \times ECF \times TCF$$

For the RMS,

$$UCP = 114 \times E0.995 \times 0.88 = 99.81$$

In the case of the RMS the overall estimated size necessary for developing software is found to be 99.81 use case points. On knowing the size of the project the effort of the project as whole can be determined.

Step 6: Estimating Effort.

$$\text{Estimated Effort} = \text{UCP} \times \text{man-hour}$$

Given that man-hours per use case point is 18

$$\text{Thus, for RMS, Estimated Effort} = 99.81 \times 18 = 1796.73$$

Estimated Effort = 1797 hours (approx.)

2.4.2 CLASS POINTS

Based on the design document, the quantification can be effectively done using a class diagram in object-oriented development. The class hierarchy and structural functionality will be contained in the specified target system. These are referred as logical blocks for the developed system. The development of the technique class point was done in 1998 (ISBSG). This class point approach is related to the function point method. By means of counting the internal attributes, the system will be denoted.

To find the measurement of a target system three main steps have been followed. These steps are illustrated in Figure 2.2. Some sort of activity must be performed in every step in order to collect the data that quantified in various classes.

The classes must be identified, and divide onto four types of system in the initial step. The different types of system are as follows:

(1) Task Management Type (TMT).
(2) Human Interaction Type (HIT).
(3) Data Management Type (DMT).
(4) Problem Domain Type (PDT).

Every type represents features of the target system.

A complex system can be easily differentiated with the help of this classification and the contrast between them can be analyzed easily. Once the finding and classification of class is completed, the level of complexity existing in each class will be determined by class points. Further, the number of attributes, the number of services requested, the number of external methods, etc., were also determined. Through using the technical complexity present in the target system, the evaluation of class points will be done. In the previous section the determination of the technical complexity factor is explained briefly. For this class point, the detailed explanation based on equation and protocol.

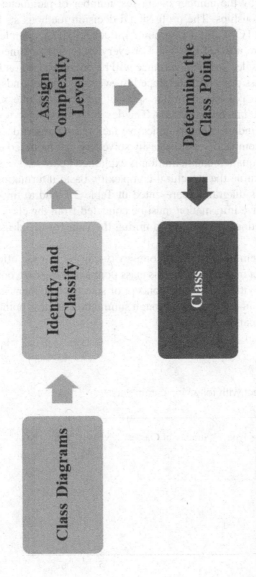

FIGURE 2.2 Class Point computation procedure.

The technique is not well suited for general software projects. The gathering of quantification information is not effectively achieved in the class point technique. Only the number of attributes, the number of services requested, and the number of external methods was calculated. Additionally, the estimating of effort in software systems is affected by the number of classes, number of parameters, and number of inheritance relationships. The professional decision methods such as Technical Complexity Factor (TCF), Total Degree of Influence (TDI), complexity level, and component type were utilized by UCP. For every factor the obtained value will be related to an expert's decision and variance will be created based on final outcome.

The modified class point was developed to overcome these kinds of issues. The developed technique possesses similar sorts of advantages, like use case points which is given in the preceding section. The developed technique will depend only on the diagram and independent of subjective factors like expert decision. The system's architectural complexity can be easily solved by this modified class point. For various factors the equation and definition is explained below.

In order to determine the structural complexity, basic information must be collected from the class diagrams represented in Table 2.7, and to find the structural complexity the relative information must be collected from the class represented in Table 2.6 from equations 7 to 13. By summing the value of the class points the CP will be obtained.

In the target system, architectural complexity can be learned effectively by the developer and project manager using this class points and use case points created on the basis of UML. Within UML the concept of size measurement may be utilized for estimating effort in the project. Through summation of class points and use case points, UML points can be obtained.

Example:

Consider the project with following parameters:

Class Type	Number of Classes	NEM	NSR	NOA
PDT	3	5	3	10
		4	5	8
		9	12	7
HIT	5	6	8	9
		7	12	10
		6	9	6
		5	7	2
		5	6	3
DMT	2	2	4	0
		1	0	0
TMT	2	4	3	0
		6	10	1

Assume all the technical factors have average influence. Estimate the effort using CP1 and CP2.

SOLUTION

The values of CP1 and CP2 are taken from Appendix A for the corresponding NEM (Number of External Methods), NSR (Number of Services Requested), and NOA (Number of attributes) given in the problem.

Class Type	Number of Classes	NEM	NSR	NOA	Complexity (CP1)	Complexity (CP2)	Weight (W1)	Weight (W2)
PDT	3	5	3	10	Average	High	6	10
		4	5	8	Average	Average	6	6
		9	12	7	Very high	High	15	10
HIT	5	6	8	9	High	High	12	12
		7	12	10	High	High	12	12
		6	9	6	High	Average	12	7
		5	7	2	High	Low	12	4
		5	6	3	High	Low	12	4
DMT	2	2	4	0	Average	Low	8	5
		1	0	0	Low	Low	5	5
TMT	2	4	3	0	Low	Low	4	4
		6	10	1	High	Low	9	4

FOR CP1

$$TUCP = \sum_{i=1}^{4}\sum_{j=1}^{3} Wij \times Xij \quad TUCP = \sum_{i=1}^{4}\sum_{j=1}^{3} Wij \times Xij$$

$$= 6 \times 2 + 15 \times 1 + 12 \times 5 + 8 \times 1 + 5 \times 1 + 4 \times 1 + 9 \times 1$$

$$= 12 + 15 + 60 + 8 + 5 + 4 + 9$$

$$= 113$$

FOR CP2

$$= 10 \times 2 + 1 \times 6 + 12 \times 2 + 7 \times 1 + 4 \times 2 + 5 \times 2 + 4 \times 2$$

$$= 20 + 6 + 24 + 7 + 8 + 10 + 7$$

$$= 82$$

$$TCF = 0.55 + \left(0.01 \times \sum_{i=1}^{18} fi\right)$$

TABLE 2.7

Mathematical Equations of Different Factors

S. No.	Factors	Definition	Equation
1.	Number of Classes (NOC)	The correlation exists between the effort estimation and quantity of classes utilized for developing the target system and then the architectural complexity of the software system will be determined.	$NOC = \sum_{i=0}^{n} noc_i$
2.	Number of Realized Relationships (NORR)	One of the correlation attributes existing among classes will be shown by this definition, then quantity of realized relationships utilized for developing the target system must be mentioned.	$NORR = \sum_{i=0}^{n} norr_i$
3.	Number of Methods (NOM)	Total number of methods utilized for developing the target system. The correlation exists between the effort estimation and quantity of classes utilized for developing the target system and then the architectural complexity of the software system will be determined.	$NOM = \sum_{i=0}^{n} nom_i$
4.	Number of Class Attributes (NOCA)	Total number of class attributes required for developing the target system,	$NOCA = \sum_{i=0}^{n} noca_i$
5.	Number of Associations (NOASSOC)	Number of associations utilized for developing target system.	$NOASS = \sum_{i=0}^{n} nass_i$
6.	Number of Use Relationships (NOUR)	One of the correlation attributes existing among classes will be shown by this definition, then quantity of use relationships utilized for developing the target system must be mentioned.	$NOUR = \sum_{i=0}^{n} nour_i$
7.	Average Number of Class Attributes per Class (ANCA_CLS)	The average number of class attributes per class within the document meant for designing.	ANCA_CLS = NOCA/NOC
8.	Average Number of Parameters per Class (ANP_CLS)	The average number of parameters per class within the target system.	ANP_CLS = NOP/NOC
9.	Number of Inheritance Relationships (NOIR)	One of the correlation attributes existing among classes will be shown in this definition, then quantity of inheritance relationships utilized for developing the target system must be mentioned.	$NOIR = \sum_{i=0}^{n} noir_i.$
10.	Average Number of Associations per Class (ANASSOC_CLS)	The average number of associations per class within target software system.	ANASS_CLS = NOASS/ NOC

(Continued)

TABLE 2.7 (CONTINUED)
Mathematical Equations of Different Factors

S. No.	Factors	Definition	Equation
11.	Class Points (CP)	The class points present in target system.	$CP = \sum NOM + $ (NOC + NOIR + NOUR + NORR + NOCA + NOASS)
12.	Average Number of Relationships per Class (ANREL_CLS)	The average number of relationships per class within the target system.	ANREL_CLS = (NOIR + NOUR + NORR)/NOC
13.	Average Number of Methods per Class (ANM_CLS)	The proportion of number of methods to class existing within target system.	ANM_CLS = NOM/NOC
14.	Number Of Parameters (NOP)	The parameters that are utilized in the specified method will be shown in this definition. The correlation exists between the effort estimation and quantity of classes utilized for developing the target system and then the architectural complexity of the software system will be determined.	$NOP = \sum_{i=0}^{n} noiri$

Given that all the technical factors have average influence. Thus,

$$TCP = 0.55 + (0.01 \times 18 \times 3)$$

$$= 0.55 + 0.54$$

$$= 1.09$$

$$CP = TUCP \times TCF$$

Thus,

$$CP1 = 113 \times 1.09$$

$$= 123.17$$

$$CP2 = 83 \times 1.09$$

$$= 90.47$$

$$Effort = 0.843 \times CP1 + 241.85$$

$$= 0.843 \times 123.17 + 241.85$$

$$= 103.83 + 241.85$$

$$= 345.68$$

$$= 346 \text{ man-month (Approx.)}$$

$$\text{Effort} = 0.912 \times CP2 + 239.75$$

$$= .913 \times 90.47 + 239.75$$

$$= 82.59 + 239.75$$

$$= 322.34$$

$$= 322 \text{ man-month (Approx.)}$$

2.4.3 THE CONSTRUCTIVE COST MODEL (COCOMO)

In his book, which is based on software economics, Boehm explained a model for estimating the cost referred as COCOMO. Generally, three stages of application will be offered by this model which includes advanced, intermediate, and basic. On the basis of factors contained in the model the advanced application will be rendered. On the basis of complexity, the application is then classified onto three modes, such as detached, semi-detached, and organic. The environment is developed using this approach. Jones (1996), explained a general effort model that is suited for every level of mode and application.

$$E = a\left(\text{EDSI}\right)^{b} \times \left(\text{EAF}\right)$$

Where,
   E = effort estimation, expressed in man-months.
   EDSI = estimate of distributed source instruction.

The determination of the parameters a and b will be done through the mode of development and the value for these parameters is found to be constant. In cases of complexity of the application the parameter is also increased. For the basic model the EAF and effort adjustment factor is found to be 1. For advanced and intermediate EAF and effort adjustment factor will be equal to a product of 15 cost factor. In terms of cost, the multiplicative cost factor is found to be reflective of a predicted proportional decrease (<1) or increase (>1). For instance, the cost factor of about 1.15 can be utilized for large degrees of reliability and the entire effort estimation is increased. If the cost factor of about 0.88 is utilized for a lower degree of reliability then the entire effort estimation is decreased (Jones, 1996).

2.5 EFFORT ESTIMATION TECHNIQUES AVAILABLE IN CBSD

There are so many effort estimation techniques available in CBSD. The following approaches are discussed:

2.5.1 PARAMETERIZED APPROACH

The main feature exhibited in CBSD is not found in the usual cost-estimation models. The consequence of scheduling will be focused at the component level.

Within the specified time in this CBSD model, the operation of several components can be carried out by a programmer and can also be done where the single component may contain many developers. In a similar time period several components can be operated by multiple developers.

The process of estimating effort carried out by this model can be analyzed in three dimensions.

Based on this view:

X-axis acquires the components which are in development.
Y-axis denotes time.
Z-axis denotes the programmer involved in the project.

The metrics and the factors that are involved in effort estimation can be effectively calculated by means of this CBSD when compared to usual models developed for effort estimation. The conception of programmer, time, and components were utilized for creating appropriate metrics. The scheduling influence in this model can be determined based on these metrics.

To summarize, the metrics are explained in Table 2.8:

The initial step will be provided by these metrics in the analysis of effort estimation through CBSD. In the field of modeling of effort the important suggestions were revealed by the research. Through this work the concept of CBSD process can be understood well and the comparison existing between the usual and CBSD models for effort estimation can be analyzed.

The pros and cons of the approach: This is a good approach to estimate effort, but the author is unable to find out the approach for CBSD. This approach does not differentiate between the effort estimation of components and the effort estimation of CBSD.

TABLE 2.8
Metrics for Effort Estimation in CBSD

Metrics	Definition
Intensity	Proportion of definite time necessary for the component to number of units of time programmed for that component.
Concurrency	The degree at which the process of unique components is performed through several developers.
Fragmentation	The degree at which the process of multiple components is performed through single developers.
Component Project Experience	Before starting the process on a particular component, the number of components finished in the respective project.
Programmer Project Experience	The programmer completes a certain number of components before allotting a specified component.
Team Size	For a particular component the total number of programmers assigned.

2.5.2   COCOMO II

A modification was made in previously developed cost estimation models and new a model for cost estimation, called COCOMO II, was developed. This model was developed for estimating the cost in recent software systems. In the COCOMO technique the primary item needed is the system component which is generated from scratch and is also referred to as the new code. However, in COCOMO II, the primary requirement is a system component created from scratch but an alteration can be made based on the necessity that is referred to as reuse code.

If COCOMO II modeling of cannot be accomplished, then the pre-existing component's source code is not accessed. The component is used as it is, only the executable file is accessed. At most, a surrounding component software shell for adopting its functionality is built. This model has three forms:

(1) Application Composition.
(2) Early Design.
(3) Post Architecture.

The Post Architecture model found in this has the direct link with Constructive Commercial Off-the-Shelf Cost Model (COCOTS).

COCOMO II builds upon the basic modeling equation. Many models related to software estimation will depend on this model because this is termed as the standard model.

$$\text{Effort} = A \left(\text{Size} \right)^{B}$$

Where

B refers to an exponential factor which deals with nonlinear diseconomies or economics of scale that might accumulate when the size of the software increases.

Size refers to the size of the software project that is to be developed (normally specified within the source code, however new measures were occasionally utilized, e.g., object points, function points, etc.

A refers to a multiplicative conversion constant that links the software program size with the development effort.

Effort refers to effort estimated in software development (generally assumed in person-months).

The Post Architecture model, as the name suggests, is typically used after the software architecture is well defined and established. It estimates the entire development life cycle of the software product and is a detailed extension of the Early Design model. This model is the closest in structure and formulation to the Intermediate COCOMO 81 and Ada COCOMO models. It uses Source Lines of Code and/or Function Points for the sizing parameter, adjusted for reuse and breakage; a set of 17 effort multipliers and a set of five scale factors that determine the economies/diseconomies. The replacement of the developed COCOMO 81 model can be made with five scale factors as well as the improvement in the Ada COCOMO model.

Example:

Suppose a system for office automation is to be designed. It is clear from the requirements that there will be four modules of size 1.5 KLOC, 2.5 KLOC, 1.0 KLOC, and 3.0 KLOC respectively. The size of the application database is high (1.08), people working of virtual m/c have low experience (0.87) and software tools are intensively used in development (0.83). All other factors are of nominal rating. Use the COCOMO model to determine overall cost and schedule estimates.

SOLUTION

The size of the project will be the sum of all four modules, thus:

$$\text{Size (in KLOC)} = 1.5 + 2.5 + 1.0 + 3.0 = 8.0$$

Since size is less than 50 KLOC, thus the mode of project will be Organic.
In the exercise, there are details about complexity factors, so to estimate the effort the intermediate COCOMO model will be applied.

$$E = a(KLOC)^b \times EAF$$

$$D = c(E)^d$$

$a = 3.2$, $b = 1.05$, $c = 2.5$, $d = 0.38$

$$E = 3.2(8)^{1.05} \times 1.08 \times 0.87 \times 0.83$$

$$= 3.2 \times 8.87 \times 1.08 \times 0.83$$

$$= 25.44 \text{ man-month}$$

$$= 25.44 \text{ man-month (Approx.)}$$

$$D = 2.5(22.14)^{0.38}$$

$$= 2.5 \times 3.25$$

$$= 8.12 \text{ months}$$

2.5.3 COCOTS

The COCOTS was developed for the purpose of performing various highlights and conditions. The following are four basic conditions of COCOTS:

(1) Calculating the candidate commercial off the shelf (COTS) component.
(2) Tailoring the COTS component.
(3) The design and evaluation of any sort of incorporation which is necessary for including the COTS component into a bigger system. This process is also referred to as "glue code".
(4) Because of volatility the level of programming and evaluation in software systems is enhanced.

Assessment: This is referred to the function in which the COTS components were chosen for utilization in the development of bigger systems.

Tailoring: This refers to the activity that is carried out for generating specified COTS programs for the purpose of utilization, the system in which it is involved will not be taken into account, even in case of a standalone operating system. The processes that come under this are setting up security protocols, specifying I/O screens or report formats, and initializing parameter values.

Glue code: The new code exterior to the COTS component will be designed and evaluated to include it in a larger system. For the specified context this determined code is distinctive within the COTS component. This process must not be correlated or confused with the above-mentioned tailoring process.

Volatility: The frequency at which the updated or newer version of COTS software is made in order to be effectively used in a larger system. The release of this larger system will be made through vendors in the event of the system's subsequent deployment and development (Construx Software, ISBSG).

2.6   FUNCTION POINTS AND OTHER SIZE METRICS: SIMILARITIES AND DIFFERENCES

Differences, as well as similarities, existing in function points as well as in the rest of the size measurement are explained as follows:

(1) The time when the calculation was performed varies. Capacity point tally was typically performed in the major phase when compared to the rest of two-size measurements: just once the realistic UI, sources of information, and yields for framework plus element correlation module was characterized. On the off-chance that capacity focuses were determined at a prior phase, real data must be used for assumptions, otherwise it may expand the estimate mistake.
(2) The quantity of components gives the least demanding approach to recognize the distinction between work focuses and different measurements. Capacity focuses have five components, but when utilizing element articles or exchange, the utilization case is the main component that is considered.
(3) The manner by which every component is evaluated. When utilizing

capacity focuses, the multifaceted nature of their components is computed in the accompanying manner:

a) The division of functional complexity for every EIF and ILF is done by means of quantity of "record element types" plus "data element types" (DET) existing within every EIF and ILF.

b) The division of function for EQ and EI is done by means of quantity of data elements type (DET) and quantity of internal and external file references (FTR).

Additionally, on utilizing the capacity point metric, the quantity of information components influences the complexity. Similar information components might be available in EIF, ILF, or EQ and EI. This could be the motivation behind why some connections are distinguished among the components of capacity focuses. It is apparent that the information components influence the intricacy of a framework even if it appears that the strategy does not accurately reflect how applicable this might be. When utilizing substance articles or exchanges the utilization case size may be determined in two elective ways: through utilizing either element items or exchanges.

(4) The idea of ILF inside the capacity focuses metric is equivalent to the idea substance object with in element object metric, however the strategies used to tally them are extraordinary.

(5) The need to utilize use case differs. Capacity focuses may be determined by utilizing use cases, however it is not mandatory to utilize them, while it is fundamental when utilizing substance items or exchanges.

EXERCISE

1. Explain the concept of use case point (UCP). It is better than FP with the reference of effort estimation? (Yes /No). Justify your answer.

2. Consider a project with the following functional units:
   - Number of external interfaces = 10.
   - Number of internal files = 12.
   - Number of external enquiries = 13.
   - Number of external outputs = 24.
   - Number of external inputs = 50.

   It is assumed that all the element weighting factors are simple and all complexity adjustment factors are average. The project is developed using C language. Estimate the effort required to develop the software.

3. If there is an additional requirement for very high level of reliability and a high volume database is needed, then what percentage of the effort will be incremented or decremented in the previous question?

4. Consider the following use case diagram:
   For the use case shown in Figure 2.3, Passenger is a complex actor while others are simple. All the use cases in which Passenger is an active actor

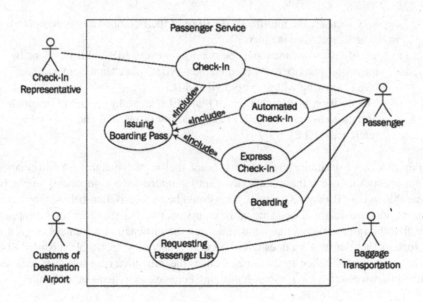

FIGURE 2.3   Use case diagram for Passenger Service System.

have average transaction and the rest use cases have simple transactions. The Assigned Value for all the Technical and Environmental Factors is 3. Calculate the Unadjusted Use Case Points (UUCP) for using the RMS. Also, calculate the total effort for developing passenger service system (PSS) if 12 man-hours per use case point will be used.

5. Consider the project with following parameters:

Class Type	Number of Classes	NEM	NSR
PDT	3	3	3
		4	6
		6	4
		8	10
HIT	5	6	8
		7	11
		4	5
		5	7
		6	7
		5	6
DMT	2	3	2
		3	3
		2	0
TMT	2	3	3
		6	8

Assume all the technical factors have average influence. Estimate the effort using CP1.

REFERENCES

Bohem, B. (1981). *Software engineering economics*. Englewood Cliffs, NJ: Prentice-Hall, Inc.

Chen, H. C. (1995). Machine learning for information-retrieval. Neural networks, symbolic learning and genetic algorithms. *Journal of the American Society for Information Science*. 46(3), 194–216.

Construx Software Inc. Retrieved from http://www.construx.com/estimate.

Fairley, Richard E. (1992). *Recent Advances in Software Estimation Techniques*. Woodland Park, CO: Management Associates.

International Software Benchmarking Standards Group (ISBSG). *Software estimation*.

Jones, C. (1996, March). Software estimating rules of thumb. *Computer*, Vol 29, pp. 116–118.

Kitchenham, B. (1997). Counterpoint: The problem with function points. *IEEE Software*, *14*(2), 29.

Shepperd, Martin, Schofield, Chris, & Kitchenham, Barbara (1996). Effort estimation using analogy. *IEEE Proceedings of ICSE*. Berlin, Germany

Smith, J. (1999). *The estimation of effort based on use cases*. Rational Software(International Software White Paper).

Somerville, I. (2001). *Software engineering* (6th ed.). Reading, MA: Addison-Wesley Publishers Limited.

3 An Introduction to Soft Computing Techniques

3.1 INTRODUCTION

A new approach called soft computing was recently developed to perform the process of computation. This technique was designed through artificial, as well natural concepts. The various approaches that come under the soft computing approach are Probabilistic Reasoning, Machine Learning, Evolutionary Computation, Support Vector Machines, Neural Network, and Fuzzy Logic. There are several advantages to soft computing, such as good tolerance of imprecision and uncertainty, cognitive ability, and strong learning. For this reason, this technology appears to be an emerging technology.

Traditional computing is hard computing, but there are certain limitations in hard computing such as approximation, partial truth, uncertainty, and imprecision. The concept of soft computing was developed to overcome these limitations. The model of soft computing was developed by mimicking the human mind.

Researcher Dr. Lotfi Zadeh (1965), developed this concept of soft computing. The technique is widely adopted in multidisciplinary fields. The main objective for developing this technique is to modify Artificial Intelligence to obtain a newer approach called Computational Intelligence. Basically, the technique of soft computing originated in 1981. Dr. Zadeh explained soft data analysis in his first work (Zadeh, 1997). After this initial description the technique of soft computing has since progressed. Through merging several fields, such as Probabilistic Computing, Genetic and Evolutionary Computing, Neuro-Computing, and Fuzzy Logic into a single multidisciplinary approach, the concept of soft computing evolved. This is the definition given for Soft Computing by Dr. Zadeh. The ultimate aim for developing this soft computing was to generate intellectual machines to solve mathematical and nonlinear problems (Zadeh, 1993, 1996, 1997).

There are two major advantages of the soft computing technique. Nonlinear problems, for which there is a lack of mathematical model, can be resolved by means of this technique. The basic knowledge that is found in human-like learning, understanding, recognition, and cognition were employed in this computing. Based on this information the generation of intellectual systems is achieved. Automated designed systems and self-tuning systems are some of the intelligent systems. The technique of soft computing is one of the newer concepts in the field of science. It has attained tremendous growth and advancement which is beyond what Dr. Zadeh, the author who initiated this concept, thought possible. For instance, fields such as Multi-Valued Logic, Evidential Reasoning, Probabilistic Computing, Evolutionary Computing, Neural Networks, Rough Sets, Fuzzy Sets, etc., which must be included in this soft computing technique, have been explained in certain literature (Kacpzyk, 2001). Within the normal soft computing concept, the introduction of Immune Network Theory and Chaos Computing was to develop a newer concept, referred to as Extended Soft Computing (ESC). The concept of ESC was developed by (Dote et al. 2000). In the previous section, soft computing and its developers were explained. The technique of ESC was developed to overcome reactive and cognitive AI, as well as to solve complex systems. Traditional probability computing techniques process soft systems computations.

The fundamental requirement for developing this soft computing was fulfilled by Fuzzy Logic. The advancement in Fuzzy Logic led to the development of Type-2 Fuzzy Logic (Mendel, 2001). Recently, the emergence of a newer science, based on Biotic and Bios Systems, is trending. The idea of Bios Computing plays a significant role in the technique of soft computing. It has been proved through many authors' expertise, who clearly explained the reason for replacing the usual computation with soft computing, and how it has been achieved.

3.2 SOFT COMPUTING TECHNIQUES

A hierarchy of soft computing techniques is shown in Figure 3.0.

3.2.1 FOUR FACTORS OF SOFT COMPUTING

It is clear from the previous section that progress and development is visualized in the field of soft computing. A proper definition of the idea enclosed in the technique of soft computing has not been known until now. Numerous ideas from various research into the technique was found. Further, many branches of science are yet to adopt soft computing. Four major components were developed by Dr. Zadeh. A detailed description based on the four factors is illustrated in Table 3.1.

3.2.2 FUZZY LOGIC

The mapping of an input space to output space utilizing the process of fuzzy interference can be achieved effectively through Fuzzy Logic. Fuzzy Logic is referred to as multi-valued logic. Among the conventional evaluation the inclusion of intermediate values can be performed (Novak et al., 1999). Words which possess indefinite meaning will be easily reasoned out and quantified using Fuzzy Logic. Using this method, problems such as ambiguous, doubtful, and contradictory opinions can be solved effectively.

Precision is considered comparatively significant for Fuzzy Logic. Through defining fuzzy sets and fuzzy numbers, imprecision in output and input variables can be found. The way of representing the knowledge that suits the concept was the definition for Fuzzy Logic, but the exact definition is not known.

This technique has been utilized to solve many problems. Cox (1995) explained that instances of Fuzziness will be seen when the specified information is vague. For instance, for the word "young", a single quantitative value is not suitable. For some people, age 27 will be young and for some other people age 20 may be considered young.

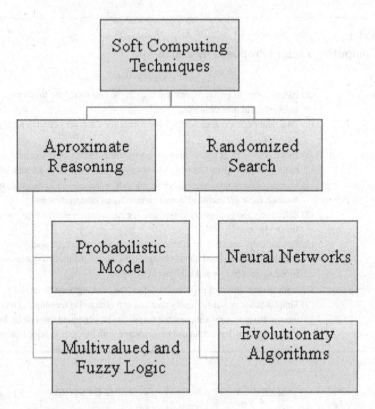

FIGURE 3.0 Hierarchy of soft computing techniques.

Why Use Fuzzy Logic?

- It is easy to understand, its concepts are very simple, and its naturalness makes it nice.
- It is flexible, it adds more functionality to any given system and is easy prevented to begin again after abrasion.
- Imprecise data can be tolerated.
- It has the ability to model nonlinear functions.
- It is based on natural language.

As shown in Figure 3.1, the creation of Fuzzy Logic is achieved by merging four main concepts which are mentioned below:

- If-Then Rules.
- Logical Operations.
- Membership Functions.
- Fuzzy Sets.

The overall fuzzy process is shown in Figure 3.2.

TABLE 3.1
Soft Computing Factors Proposed by Zadeh

Factor	Explanation
Premises	(1) Extensively, the problems that are exist in the real world are found to be indefinite and indeterminate.
	(2) Two major criteria necessary for improving the costs are certainty and precision.
Principles	(1) Drawbacks such as approximation, partial truth, uncertainty, and imprecision must be overcome for attaining low solution cost, robustness, and tractability.
Implications	(1) In spite of employing the FL, SVM, and NN techniques in a competitive manner, these are included in soft computing as complementary.
	(2) Effectively merged specified systems. for instances is referred to as "neuro-fuzzy systems".
	(3) Based on the needs of the customer, such as camcorders, photocopiers, washing machines, and air conditioners, the above-mentioned system was found to be effective and is increasing.
Unique Property	(1) With the assistance of experimental data, the learning must be attained.
	(2) Simplification power is already seen is a soft computing technique. Through interpolating or approximation the output can be generated from input. In cases of unseen input, the output can be generated with the help of prior-learned input.

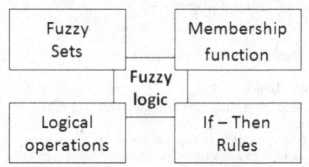

FIGURE 3.1   Fuzzy Logic concepts.

Adaptive Neuro-Fuzzy Inference System (ANFIS)

The technique which is explained in the previous section was used and verified experimentally and theoretically in various kinds of applications. But this technique is not suitable for biological counterparts (Jain and Martin, 1998). The merging of these techniques into a single multidisciplinary field was carried out by Lotfi Zadeh which led to the development of soft computing techniques.

Many further techniques were added, along with the technique of soft computing, which paved way for the development of Extended Soft Computing. The utilization of any of these techniques comes under the class of soft computing.

For the purpose of tuning the fuzzy logic (FL) controller, the neural network artificial neural network (ANN) was utilized. Lee and Lee (1974) developed a model

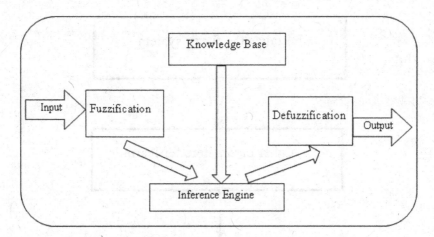

FIGURE 3.2   Fuzzy process.

based on neurons in which multi-inputs and multi-outputs can be included. The widely used normal method will rely on a binary system for obtaining output. Many researchers were motivated to carry out their research in the area of neural networks because many issues related to this network were mentioned in IEEE Communications Magazine. To solve issues in neural networks, fuzzy-based neural networks were developed (Plevyak, 1992). Following that, Jang developed the technique of Adaptive Neural Fuzzy Inference Systems (ANFIS) (Jang, 1993). In the Department of Electrical and Computer Engineering of the Institution Superior Técnico, Lisbon, Portugal in 1998, Dente and Costa Branco developed an electro-hydraulic system based on the Neuro-Fuzzy approach.

The similar kind of operation seen in neural networks is found in the neuro-adaptive learning method. The modeling protocol related to fuzzy will be rendered through neuro-adaptive learning techniques. This process is performed to obtain the information related to the dataset. The toolbox function named ANFIS, with the help of knowing the dataset, either inputs or outputs the construction of a Fuzzy Inference System (FIS). Parameters such as a membership function in this fuzzy system will be altered utilizing the least squares method or back propagation algorithm. The alteration made will support the fuzzy system to acquire the information regarding the dataset used for modeling. The shaping of the membership function can be done through the Neuro-Fuzzy Design application. Instead of giving the data manually the data which are to be used as input can be trained using the above-mentioned application. The working principle for ANFIS is given in Figure 3.3. The objectives of ANFIS are mentioned as follows:

From Neural Networks (NN) and Fuzzy Systems (FS) the integration of effective features:

(1) Integrating feature from FS: The information that is obtained previously is defined in a set of constraints in order to decrease the search process carried out through optimization.

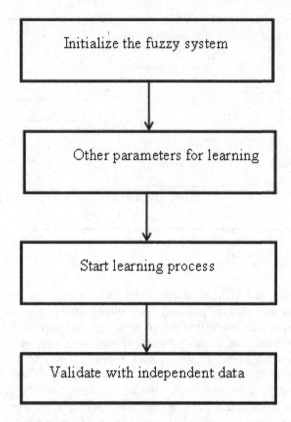

FIGURE 3.3   Basic flow diagram of ANFIS computation.

(2) Integrating feature from NN: To achieve the adjustment of parameters in FC automatically, the back propagation algorithm is utilized along with the structured network.

In certain manufacturing the ANFIS can be applied to:

(1) Models (to describe earlier data and to guess future behavior).
(2) Controllers (automated FC tuning).

Basic flow diagram of computations in ANFIS is shown in Figure 3.3.

Fuzzy Logic Toolbox

The above-mentioned Fuzzy Logic Toolbox™ is referred to as an application. Through this application a Simulink® block, application and function can be rendered. With the help of the Simulink block the process of simulation, designing, and analyzing can be performed related to fuzzy logic. The steps required for constructing fuzzy inference systems will be explained through this application. This application is considered the origin of many techniques such as adaptive neuro-fuzzy learning and fuzzy clustering. By means of a simple logic rule the modeling of complex systems can be done using this toolbox and further, the implementation of the rules can be done in a fuzzy inference system. This toolbox can be utilized as standalone fuzzy inference engine.

Key Features of the Fuzzy Logic Toolbox

- The inference systems can be constructed as well as the analysis of outcomes.
- Membership functions which are necessary for generating fuzzy inference systems can be developed.
- The logic gates such as NOT, OR, and AND logic can be defined in user-defined rules.
- Mamdani which is found to be the standard for FIS as well as Sugeno-type is considered as standard for ANFIS fuzzy inference systems.
- The alteration in membership function can be done automatically utilizing fuzzy-clustering and neuro-adaptive learning techniques.
- It has the capability to create standalone executable fuzzy inference engines or embeddable C code.

3.3   EVOLUTIONARY ALGORITHMS

The evolutionary algorithm is one of the various prevailing meta-heuristic algorithms. The evolutionary algorithm was found to be fundamental because it was developed in the past and it can be used for solving many optimization problems. To achieve efficacy in searches various parameters have been defined. Many of the rough guidelines related to empirical experience is present but the corresponding setting is complex to find. For solving these sorts of issues non-static parameter control can be utilized. Three kinds of non-static parameter controls are present which are classified by Bäck (1996). The modest variant is found to be dynamic parameter control. Only on the basis of certain schemes will the parameter be set. These schemes will mainly be influenced by quantity of generation. The scheme of control is considered significant in the case of adaptive parameter controls. The single encounter is taken as a function value. Some of the parameters that are found in the evolutionary algorithm such as selection, crossover, and mutation must be utilized for creating the parameter to be used in self-adaptive parameter control. The three above-mentioned variants are utilized now, but there is a theoretical model related to it. For the case of discrete objective function-related optimization, these variants are well suited. The continuous theoretical model was developed in evolution approach (Schwefel, 1995; Beyer, 1996; Rudolph, 1997b).

3.4   APPLICABILITY OF SOFT COMPUTING
TECHNIQUES IN SOFTWARE ENGINEERING

The technique called soft computing has been utilized broadly recently in the solution of software engineering problems. Different soft computing techniques are applied in different ways within software business. Mainly, the applicability of techniques related to soft computing in the field of software business can be classified into following four classes:

- Neural Network Concepts Usage in Software Engineering.
- Fuzzy Logic Concepts Usage in Software Engineering.
- Genetic Algorithm Concepts Usage in Software Engineering.
- Machine Learning, Deep Learning, and SVM Concepts Usage in Software Engineering.

3.4.1 FUZZY LOGIC CONCEPTS USAGE IN SOFTWARE ENGINEERING

Utilizing the estimation models in the field of software engineering, the prediction of significant attributes in future entities can be achieved, for example, productivity, software development effort, software development cost, and software reliability.

The management of software-related projects must be carried out accurately and the prediction and scheduling of effort estimation must be done based on the time to maintain the software system effectively. This process is considered as crucial in software systems.

There are so many techniques for estimation prediction, among which analogy-related estimation is a commonly utilized approach by researchers as well as industry, especially in software effort estimation. In case of analogy-related software effort estimation there are limitations. The categorical data (nominal or ordinal scale) is usually used for describing software projects, such as, very high, high, low, and very low. The attributes that are referred in this fuzzy logic are referred to as linguistic values. The challenge is faced by the concept of measurement in the software business, because of two major key motives: lack of standards, and most of the time the nature of software attributes is qualitative instead of quantitative.

Thus, at present there is a lack of a standardized estimation model to estimate software cost, effort, productivity, etc. It is still a challenge in the field of software work. Even though research is continuously carried out the estimation model is compromised with precision and certainty of value.

Soft computing might be the best solution for such problems since it has the features like the ability to work with uncertainty, learning from experience, and is tolerant of imprecise information that are likewise the key issues of software estimations.

Fuzzy set theory principle was first discussed by Zadeh. An approach that is well-suited for linguistic values was found to be the fuzzy set model principle. The creation of a quantitative framework is the main objective for fuzzy set theory. An unclear gathering of information can be limited by means of this model. The information acquired will be usually described in the form of natural language.

We can join the benefits of fuzzy logic with key challenges of software effort estimation. In recent studies it is found that for describing the categorical data the fuzzy set is found to be effective when compared to other classical intervals (Idri et al. 2000, 2001).

3.4.2 ARTIFICIAL NEURAL NETWORK (ANN) CONCEPTS USAGE IN SOFTWARE ENGINEERING

In the field of software engineering, the estimation of effort is one of the main, as well as critical, tasks. So many times, the success of the project depends on the accuracy of effort estimation. Not only underestimation, but overestimation also creates lots of issues. Overestimation can result in the loss of bids and underestimation can result in financial loss for the company.

It was proved that about seventy to eighty percent of projects based on software are not issued with the correct declared costs, according to Standish Group International and the International Society of Parametric Analysis (ISPA) (Eck et. al., 2009). There might be a number of reasons for software failure but the leading factors are system requirements, inappropriate team size, uncertainty of software, inadequate estimation of project cost and size. The accuracy of estimated effort depends on the precision of project size expected.

High precision in estimating effort is still a big challenge for software developers as well as the research community. Although research is ongoing, and various mathematical models like regression analysis, COCOMO, and Regression Tree are proposed by researchers, it is not sufficient.

There is a need to maximize the accuracy of prediction and it could be achieved by introducing the concepts of artificial neural network in effort estimation. Among the various supervised learning techniques, Artificial Neural Network (ANN) is among them. Many neurons, referred to as elementary processing units, will be linked to one another to create the neural network. Neural computing will be carried out with the help of this neural network. The arrangement of neurons will be found in different layers and the data which is used as input will be within the input layer. By means of this network the data will be transferred to the output layer where the result will be obtained. The neurons inside the network are referred to as processing units. Many units of input can be processed by this neuron to get the desired output. The weight will be linked with every neuron represented as input in order to alter the strength of the input.

3.4.3 GENETIC ALGORITHM CONCEPTS USAGE IN SOFTWARE ENGINEERING

Recently, Genetic Algorithms (GAs) have been utilized in handling numerous product design issues. The GAs have been utilized in different periods of programming advancement, like from the necessity and investigation stage and programming testing stage. It is additionally used for growing new measurements. This will help developing programming design control, so there is a need to use GA ideas to tackle program-building discipline issues, looked at by programming experts.

At present, agile-oriented software development is the prime choice of the developers for many reasons. Quick changes in the requirements, demand for fast delivery of the product are a few examples. There are two major challenges in such a scenario, one is to identify/select the best software components from the repository for software development and other is to reduce the quantity of test cases utilized for testing.

Programming testing is the way toward executing a program with the goal of discovering bugs. Programming testing expends significant assets in terms of exertion and time in programming an item's lifecycle. Experiments and the age of test information is the key issue in programming testing and its robotization improves the proficiency and viability and brings down the significant expense of programming testing. The age of test information utilizing dynamic, emblematic, and irregular methodology is not sufficient to produce an ideal measure of test information. Some different issues, similar to no acknowledgment of events, of endless circles, and wastefulness to create test information for complex projects, makes these methods unacceptable for producing test information.

The other challenges in software development are software project effort estimation, software testing, quality measures, reliability, and maintenance. In the field of software development, the concept of estimating the software cost is found to be a significant and challenging task. Many techniques have been developed to perform the estimation of cost effectively but none of the methods work effectively.

3.4.4 SUPPORT VECTOR MACHINE (SVM) CONCEPTS USAGE IN SOFTWARE ENGINEERING

Programming failures are costly in terms of quality and cost. Additionally, the expense of identifying and remedying failures is a pricey programming action. To limit the expense of the product advancement and to improve the adequacy of the product testing process, a module's deficiency inclination estimation of is significant movement. Additionally, to create a subjective item it is fundamental to foresee deficiencies proactively. A testing procedure can be improved by early expectations of imperfections. Along these lines, Software Defect Prediction (SDP) is basic in program building. In SDP, the modules might be sorted into two classes, damaged and non-damaged. Order is a managed-learning approach. Along these lines SDP is a regulated twofold arrangement issue. Since Support Vector Machine (SVM) is a propelled-order technique, along these lines it is the best methodology for deserting arrangements. SVM has numerous focal points like giving a worldwide information order arrangement. SVM is utilized by scientists to build a prescient SDP model (Sundararaghavan and Zabaras, 2005). The following are some driving arrangements developed by different scientists for programming building issues utilizing SVM:

(1) For predicting software defect numbers, a new method utilizing Fuzzy Support Vector Regression (FSVR) was developed by Yan et al. (2010).
(2) For predicting errors in new software product version, a Twin Support Vector Machine (TSVM) was presented by Agarwal and Tomar (2014).
(3) Shan et al., presented a novel model related to Locally Linear Embedding and SVM (LLE-SVM) (Shan et al., 2014). The software security can be enhanced through SDP by serving testers for finding the software error more exactly.

EXERCISE

(1) What kind of benefits can be achieved by an organization through the usage of fuzzy concepts in software cost prediction over the COCOMO?

(2) How does early defect prediction support software quality?

(3) In software effort estimation, ANN and fuzzy concepts can be used to achieve highly accurate predictions. Suggest some scenarios where ANN could be better than the Fuzzy approach and vice versa.

(4) What are the possible situations when GA concepts cannot be used to improve software testing quality?

(5) Why is SVM a better solution for SDP?

REFERENCES

Agarwal, S., & Tomar, D. (2014, March). Prediction of software defects using Twin Support Vector Machine. In *IEEE International Conference on Information Systems and Computer Networks (ISCON), 2014* (pp. 128–132) Mathura, India.

Bäck, T. (1996). *Evolutionary algorithms in theory and practice.* New York, NY: Oxford University Press.

Beyer, H.-G. (1996). Toward a theory of evolution strategies: Self-adaptation. *Evolutionary Computation, 3*(3), 311–347.

Costa branco, P. J., & Dente, J. A. (1998).An Experiment in automatic modelling and electrical drive system using fuzzy logic. *IEEE Transaction on Systems Man and Cybermetics, 282,* 254–262.

Cox, E. (1995). *The fuzzy systems handbook.* AP Professional, Buffalo, New York.

Dote, Y., Taniguchi, S., & Nakane, T. (2000). Intelligent control using soft computing. In *Proceedings of the 5th Online World Conference on Soft Computing in Industrial Applications (WSC5),* Helsiniki, Finland.

Eck, D., Brundick, B., Fettig, T., Dechoretz, J., & Ugljesa, J. (2009). *Parametric estimating handbook* (4th ed.). The International Society of Parametric Analysis (ISPA), Chandler, Arizona.

Idri, A., Abran, A., & Khoshgoftaar, T. M. (2001). Fuzzy Analogy: A new Approach for Software Cost Estimation. In *Proceedings of the International Workshop on Software Measurements* (pp. 93–101). Montreal, Canada.

Idri, A., & Abran, A. (2000). Towards A fuzzy logic based measures for software project similarity. In *Proceedings of the 61 Maghrebian Conference on Computer Sciences* (pp. 9–18). Fes, Morocco.

Idri, A., & Abran, A. (2001a). A fuzzy logic based measures for software project similarity: Validation and possible improvements. In *Proceedings of the International Symposium on Software Metrics* (pp. 85–96). England, UK: IEEE Publications Computer Society.

Idri, A., & Abran, A. (2001b). Evaluating software projects similarity by using linguistic quantifier guided aggregations. In *Proceedings of the 6th IFSA World Congress and 2(/h NAFIPS International Conference* (pp. 416–421). Vancouver, Canada.

Idri, A., Kjiri, L., & Abran, A. (2000). COCOMO cost model using fuzzy logic. In *Proceedings of the International Conference on Fuzzy Theory and Technology* (pp. 219–223). Atlantic City, NJ.

Jain, Lakhmi C., & Martin, N. M. (1998). *Fusion of neural networks, fuzzy systems and genetic algorithms: Industrial applications.* Boca Raton, FL: CRC Press.

Jang, J. R. (1993). ANFIS: Adaptive-Network-based fuzzy inference system. *IEEE Transaction on Systems, Man, and Cybernetics, 23*(3), 665–685.

Kacpzyk, Janusz (Eds.). (2001). *Advances in soft computing*. Heidelberg, Germany: Springer-Verlag.

Kacpzyk, Janusz (Eds.). (2001). *Studies in fuzziness and soft computing*. Heidelberg, Germany: Springer-Verlag.

Lee, S. C., & Lee, E. T. (1974). Fuzzy sets and neural networks. *Journal of Cybernetics*, *4*(2), 83.

Mendel, Jerry (2001). *Uncertain rule-based fuzzy logic systems: Introduction and new directions*. Upper Saddle River, NJ: Prentice Hall.

Novák , Perfilieva I., & Mockof, J. (1999). *Mathematical principles of fuzzy logic*. Dodrecht: Kluwer Academic Publishers.

Plevyak, Thomas (Eds.). (1992). *IEEE Communications Magazine*. (Special Issue: Fuzzy and Neural Networks), *30*(9).

Rudolph, G. (1997b). How mutation and selection solve long-path problems in polynomial expected time. *Evolutionary Computation*, *4*(2), 195–205.

Schwefel, H.-P. (1995). *Evolution and optimum seeking*. New York, NY: Wiley.

Shan, C., Chen, B., Hu, C., Xue, J., & Li, N. (2014, May). Software defect prediction model based on LLE and SVM. In *Communications Security Conference (CSC 2014)* (pp. 1–5). IET, Beijing, China.

Sundararaghavan, V., & Zabaras, N. (2005). Classification and reconstruction of three-dimensional microstructures using support vector machines. *Computational Materials Science*, *32*(2), 223–239.

Yan, Z., Chen, X., & Guo, P. (2010). Software defect prediction using fuzzy support vector regression. In *Advances in Neural Networks-ISNN 2010* (pp. 17–24). Berlin Heidelberg: Springer.

Zadeh, L. A. (1965). Fuzzy sets. *Information and Control*, *8*(3), 338–353.

Zadeh, L. A. (1975). The concept of a linguistic variable and its application to approximate reasoning. *Information Sciences* Part I: 8, pp. 199–249; Part II: 8, pp. 301–357; Part III: 9, pp. 43–80.

Zadeh, Lotfi (1993). Fuzzy Logic and Soft computing, Plenary Speaker. In *Proceedings of the IEEE International Workshop on Neuro Fuzzy Control*, Muroran, Japan.

Zadeh, Lotfi (1996). The role of soft computing and fuzzy logic in the conception, design, development of intelligent systems, plenary Speaker. In *Proceedings of the International Workshop on soft Computing Industry*, Muroran, Japan.

Zadeh, Lotfi (1997). *What is Soft Computing, Soft Computing*. Germany/USA: Springer-Verlag.

4 Fuzzy Logic-Based Approaches for Estimating Efforts Invested in Component Selection

4.1 INTRODUCTION

Every developed software company must play a significant role in Software Effort Estimation. Effort estimation is one of the most important and most analyzed tasks that a developer has to face. Component-Based Software Engineering (CBSE) is the new area in software development. The source of origin and features for these two component-based software and Commercial off the Shelf (COTS) software components are different, but on combining these components an effective software system may be developed. The concept of designing software related to components is significant nowadays, especially in the software industry. Based on this technique the combination of these components can help in attaining an effective application. The achievement of effective application relies on the selection of components. It is very necessary to evaluate the efforts invested for selecting these components. However, the market is working quickly in terms of utilization of these approaches, but a lower number may be achieved by means of the measurement of efforts invested in this emerging approach.

Software Estimation models can be classified into three groups:

Expert judgment: These methods are based upon the experts. Experts in this field may provide judgment easily without using any complicated methods. However, because it is a human-dependent method it will be difficult to repeat.

Comparison-based methods: These methods may be used as good alternative to expert-based methods. In these estimation methods first a similar project in history is identified, and then on the basis of that project, estimates may be derived. These methods may be considered as a systematic approach to expert methods.

Algorithmic models: These methods are known to be the most popular models. In these methods using statistical data. an analysis parametric equation for estimation is established, e.g., COCOMO, SLIM.

Fuzzy set theory provides a framework for addressing the uncertainty of many aspects of human cognition. Nowadays, fuzzy governance-based systems are applied to a wide variety of real-world problems from a variety of sectors and to many real-world applications, such as stability or multi-criteria decision domains. Although a system may be defined mathematically in general terms, engineers and researchers still prefer fuzzy logic system representation. The application of fuzzy logic is under investigation due to the inherent uncertainty of software estimation. Thus, to include qualitative and quantitative aspects of software selection projects, the project proposes a goal-oriented programming model for optimization of the selection process. Such studies are meaningful but do not rely directly on software assessment. The regression model is one of the options for software assessment and proposes a regression model for software estimation, enhanced by the application of a fragile fuzzy injection system.

Several methods are proposed for estimating efforts using fuzzy logic; these methods are based upon some factors or parameters. Some fuzzy logic-based proposed models for efforts estimation of CBSD are as follows:

Martin et al. (2005) describes an application comparing multiple regression. Based on ten programs, a subset consisting of 41 modules was designed and it was considered as data. From the experimental result it was found that the utilization of fuzzy logic mean magnitude of relative error (MMRE) achieves higher result when compared to using multiple regression of the MMRE. The value for (20) was found to be higher only after using fuzzy logic. The higher value of (20) was not achieved using multiple regression. Additionally, for about six of the 41 MREs subset the value of zero was obtained on using fuzzy logic. (On using multiple analogous models the same sort of case cannot be illustrated).

The analysis is carried out in this chapter for a single level by means of small programs. Then the results were calculated utilizing simple multiple regression and a fuzzy logic system. From the experimental result it was found that for designing software fuzzy logic can be used as a substituent. The research was carried out to evaluate the software expansion efforts on the rest of the other modules on which fuzzy logic systems can be executed. A high rate of coefficient (r2) is given for the data subset. Much research was carried out for combining multiple regression and fuzzy logic through which the lower rate of coefficient r2 can be achieved. Additionally, to obtain a hybrid system the research must be focused for correlating artificial neural networks along with fuzzy logic.

Carrasco et al. (2012) describes that estimation of effort in software design is found to be significant. So, for improving the accuracy in this estimation the framework, namely the SEffEst (Software Effort Estimation) framework, was developed in this present work. In this framework the combination of a neural network and fuzzy logic is used. The training phase in this framework is carried out through

ISBSG data and the best result is achieved with a higher rate of prediction accuracy. The authors tested the same framework using ANN, but without the fuzzy in these regression problems it is necessary to estimate the value from a set of input variables. Previous research work showed that the hit rate performed by 24 authors was 0.8525. This implies that SEffEst improves the hit rate by 6.15%. The reason for this improvement is the SEffEst fuzzy logic component, which allowed for the inclusion of some test cases in which some information was vaguely determined. Respectfully, although the domain is not comparable (software predictions vs. ballistic effects), it can be seen that the SEffEst hit rate is similar to the two scenarios proposed. First, all hit rates can be considered too high, commenting on the good results SEffEst has achieved. Finally, although the research functions are different, they depend on the development of the optimization method and its implementation in the concrete framework. Therefore, it can be assumed that the optimization method can be applied to different domains that achieve excellent results. It can also be concluded that the inclusion of new elements in the methodology (such as fuzzy logic components in SEffEst) allowed for improved results. Research results indicated that the ANN structure was obtained for accurate fuzzy logic – effort estimation. Through utilizing SEffEst the mean correlation that was achieved was 0.9140. This implies that compared to previous work nearly 6.15% improvement was achieved. For the case of estimation of software designing effort the value that was obtained was found to be satisfactory. This estimation of effort mainly depends upon a greater number of variables. Future research will focus on the application of SEffEst in project phases where more variables are known. To play with more relevant and accurate information about the project, provides a better perspective.

Xu and Khoshgoftaar (2004) proposed a technique based on the fuzzy model. The language data can be performed by means of a technique called Innovative Fuzzy Recognition Cost Estimation Modeling. Through this approach the membership rules and function can be developed spontaneously. Comparison analysis is made by the proposed method with another three models related to COCOMO such as extended, intermediate, and basic by means of utilizing the COCOMO 81 database. From the comparison analysis it was found that better a cost was achieved by means of the fuzzy method in contrast with to other three methods. The finding of effort estimates and software costs is mostly done through the COCOMO cost estimate model. In this model the dataset is created from many software developments. The estimation of effort and cost will be carried out with the help of this dataset. In cases of software development, the size and difficulty increases, due to which the desired accuracy is not possible to achieve in estimation. The software development that is used as fuzzy recognition software to perform the cost assessment technique will contain some unclear information. The developed fuzzy-related technology for the purpose of prediction is the combination of several techniques such as diffusion, fuzzy effects, space projection, and fuzzy clustering. The model developed based on fuzzy is found to be quite simple and related to the fuzzy group number, the set of infection rules will be provided as similar. The preprocessing technique which is carried out related to data-driven will lead to a decrease in the size of the database.

The input data will be provided by means of fuzzy in this present work for obtaining the membership function and rules. The comparison analysis is carried out for the cost estimation of the proposed method and the other three models such as detailed, intermediate, and basic COCOMO models. From the experimental result it was proved that a better accuracy rate is obtained by utilizing a fuzzy model in comparison to the Kokomo model. Other issues, like the size of the project, can be overcome by means of fuzzy model in the upcoming research. The proposed fuzzy model can be analyzed through several case studies.

Prakash, Mittal, and Mittal (2010) implemented fuzzy logic for SCE (Software Cost Estimation). Authors described cost estimation models as one of the hot topics for the software development industry. Fourteen projects are used in this research (which includes KEMERER). The results show that the mean absolute error percentage of relative error and productivity rate improved compared to algorithmic methods. The number of application lines and application thousand-lines have a direct impact on software cost estimation.

Lalit Patil et al. (2014) proposed a technique related to quantitative measurement will be compared for effectively achieving accuracy in effort estimation within software development. This is the main purpose of this chapter. Naturally, for estimating the size more accurately in Component Point Component the method called Black Box is well suited. Secondly the comparison of the proposed model based on fuzzy logic is used with the COCOMO II model for finding the input, as well as to find newer cost driver estimation methods, in order to achieve accuracy in estimation of effort.

The analysis must be carried out for newer cost driver estimation in future in order to achieve effective outcomes. This analysis will be carried out by software companies only after obtaining many components related to software.

Some important parameters proposed for software estimation efforts are described in Table 4.1.

In this chapter a model proposed by Seth et al. (2009) that integrates five factors, namely Reusability, Portability, Functionality, Security, and Performance, and provides a measure of Component Selection efforts, is explained.

4.2   FACTORS AFFECTING COMPONENT SELECTION EFFORTS

The factors that have a great effect on Component Selection Efforts in the proposed method are explained in this section.

4.2.1   Reusability

The software component C can be used along with the product P, but the motive for designing C is not to use it along with P. This is called software reuse in Component Related Development. The method of using one major component in other setting is given as another definition. There are certain advantages in reusing software, which include the enhancement of software quality, maintainability, and productivity. In

TABLE 4.1
Software Effort Estimation Parameters

S. No.	Parameters	Definition
1	Intensity	This is the ratio of the quantity of actual time spent on selecting components to the number of time units scheduled for the component.$$\text{Intensity} = \frac{\text{Actual time spent on a component}}{\text{Total number of time units scheduled}}$$
2	Concurrency	Number of multiple programmers working simultaneously on a single component. For example, if n number of programmers are working on a component at time t then concurrency will be n.
3	Fragmentation	The degree to which a single programmer is working simultaneously on multiple components.
4	Component Project Experience	The number of components that have been completed as part of the project prior to work beginning on a particular component.
5	Programmer Project Experience	The number of components that have been previously completed by the programmers assigned to a particular component.
6	Team Size	The number of programmers assigned to a particular component.
7	Modularity (MD)	The division of software components onto other products or smaller modules to a certain level is referred to as Modularity. The service rendered by this modularity will not depend on another service. The modularization of the function must be done effectively for the purpose of becoming a reusable service. More modularity must be achieved in order to attain better reusability. The modularity must exist in the range of 0 to 1. In the present work, the modularity is divided into more, medium, and less categories.
8	Interface Complexity (IC)	With the aid of a distinct interface, the collaboration among software component and application will be achieved. In the software, the principal basis for maintenance, implementation, and coding can be rendered through this interaction. For certain sorts of components, interface complexity becomes tedious for making it understandable and. For the case of lower effort, interface complexity is found to be less.
9	Maintainability (MN)	Maintainability means that a software product or module can be customized while retaining its predefined functionality. As a result, errors can be corrected and lead to higher software quality and performance. For good reuse, maintenance must be high.
10	Flexibility (FX)	Flexibility and reusability are is the level of adaptability to a software product or component that can be received when external changes occur. For reusable components, less effort is required to modify the operating code or program, as fewer software components or products are required to analyze, design, code, and test. For good reuse, flexibility must be high.

(Continued)

TABLE 4.1 (CONTINUED)
Software Effort Estimation Parameters

S. No.	Parameters	Definition
11	Adaptability (AD)	Adaptability is the ability of the software component to adapt to new and different platforms and technologies. To reuse software extensively, one component must be compatible with different platforms. For better reuse, compatibility should always be as high as possible.
12	Relative Error (RE)	The RE is used to calculate the estimation accuracy. $$RE = \frac{\left(\text{Estimated Efforts} - \text{Actual Efforts}\right)}{\text{Actual Efforts}}$$
13	Magnitude of Relative Error (MRE)	$$MRE = \frac{\left\vert\text{Estimated Efforts} - \text{Actual Efforts}\right\vert}{\text{Actual Efforts}}$$
14	Mean Magnitude of Relative Error (MMRE)	The MMRE calculates the mean for the sum of the MRE of n projects. Specifically, it is used to evaluate the prediction performance of an estimation model. $$MMRE = \frac{\sum_{i=1}^{n} MREi}{n}$$
15	Prediction Level (PRED)	$$PRED(L) = \frac{k}{n}$$ Where k denotes the number of projects in a set of n projects, n denotes the total number of projects and L denotes the maximum MRE for a selected range, where MRE $\leq$ L, the ratio of projects' MREs which lie within the selected range (L) out of the total projects is calculated by PRED. (e.g. $n = 100$, $k = 80$, where L = MRE $\leq$30%: PRED (30%) = 80/100 = 80%).
16	Cyclomatic Complexity	A software metric provides a quantitative measure of the logical complexity of a program module. It defines the number of independent paths in the original set of the program. Any path through the independent path program introduces a new set of minimized processing statements or a new condition. Simple calculations and complexities can be calculated when calculating one.

the form of boxes, the components that are reused can be utilized. The boxes are classified as follows:

Black box reuse: The execution of the component will not be shown to the reuser, only the interface will be visualized. Within the interface of the component, restrictions, requirements, user documentation, and public methods will be considered. In cases of altering the code in the black box component, information by means of linking and compiling will be sent to the component which is to be reused.

Glass box reuse: It is possible checked the components present both outside and inside of the box using this method. However, this method is not suited for touching the component inside. The component in the box can be understood by re-user in this method. This is the advantage of this method when compared to black box reuse.

White box reuse: The interface, as well as the component in the box, can viewed and altered in cases of white box reuse. By means of delegation or inheritance the execution of specified components in a box can be shared with another box in this white box.

In Component-Related Systems reusability is an intrinsic property. The process of using the specified component in other circumstances is referred to as reuse. Among the various existing properties of Component-Based Development (CBD) reusability is a fundamental topic. The component which is to be reused should be generic, so that appropriate features are enabled for the reuse to generate the instances of the component in order to suit the reuse application. Reusability is considered a factor in the proposed method. In cases of similar types of component in larger numbers, within this existing component the suitable component will be reused many times. The selection of components does not have a major impact on the effort estimation. This implies that reusability is inversely proportional to the effort estimation.

$$\text{Reusability} \propto 1/\text{Selection Efforts}$$

The problem arises in the selection of the best appropriate component that is well suited for reusability. Basically, some protocols were used for the selection of the reusability component, such as good documentation, simple structure, and small size of code.

4.2.2 Portability

Transferring components from one environment to another is referred to as portability. This portability is also considered an important factor in the proposed system. During transfer the necessary alteration can be made. The transfer of components must be done quickly and easily to the appropriate surroundings. The time and cost adopted for portability must be reduced. This portable factor is considered significant in the case of this Component-Related Development because the use and reuse of the component will be in varying environments. There are certain components which do not depend upon the platform. In this situation portability can be easily achieved. This implies the component to be transferred must not depend on the platform. On the site www.jars.com the components that are found seem to be easily portable because the surroundings do not require any alteration and can be used effectively in various sorts of application. But for certain components, if they get transferred from one environment to other, alteration is must. Due to these reasons the of portability factor is considered as significant in Component-Related Development. This implies that in estimation of component selection efforts, portability can be considered an important factor. The selection seems to be larger if the portable nature of the component is high.

4.2.3 FUNCTIONALITY

Functionality is based on the number of functions and their related properties. The specific function will be offered by the component. Based on this function the process must be carried out by the component under certain circumstances. Already existing components will offer low cost and faster delivery. The selection of effort is high if the functionality is found to be higher.

4.2.4 SECURITY

The information that is transmitted, processed, stored, and created by the software must be secured and is the main objective in software; its availability, integrity, and confidentiality must be preserved. The executed program, as well as the resources, must be protected. The secured software has high probability because the data will be safe from any sort of modification or unauthorized access and the information will be in a stable condition A high rate of guarantee is necessary in certain sorts of application. Limitations in security have a significant effect on software in two major ways. One is functional security. In this the size of the function that is created in software is increased. The necessity of this function must be considered along with all other functional necessities and must be cleared on using homegrown code or COTS components. Another is nonfunctional security. In this, additional processes is required to achieve a certain level of guarantee in security verifications, testing, and documentation. The black box products will mostly create COTS components. These black boxes are referred to as third parties. Within an enterprise's information system utilization of this component will create significant potential risk affecting reliability and security. For instance, if certain organizations use the internet through COTS components the data stored can be leaked to the global network.

In cases of accessing unauthorized services or resources by the component: For instance, the past file can be read by the attacker by means of sent mail and share it with an external user.

Unauthorized access to the resource by the component: The alternative component may also fail in this approach. For instance, the web server which is broken in an attempt to write in a HTML file, then either only reading can be carried out or else overwriting on the home page can be done.

Certain authorized rights may be misused by the component: For instance, the root privilege can be given as a command by an attacker in order to obtain full control of the system.

The ability of the component to protect any unauthorized access to the services is referred to as security. The estimation of effort in component selection will be gradually lowered if enhanced security is found in the component.

4.2.5 PERFORMANCE

The capability of the component to render its specified function is referred to as the performance parameter. Any lack of performance is because of component technology. The main reason is that the run time system uses a resource which lacks in performance. Through an interaction mechanism the performance is also affected. Optimization can be carried out within the component to enhance the performance, but during this process the appropriate feature must not be affected. To evaluate the performance, the testing of the component needs to be carried out in various platforms. The component exhibits two types of behavior. To evaluate the performance these behaviors have to be tested. The two behaviors of the component are mentioned as follows.

Time behavior: In the appropriate circumstances, the capability of the component to carry out particular tasks at the right time.

Resource behavior: In the specified circumstances the quantity of resources utilized is referred to as resource behavior.

Through the values obtained from the parameters the ability of performance and how accurate it is can be analyzed. As performance increases the component selection effort is gradually decreased.

4.3 FUZZY LOGIC

A researcher called Zadeh developed the fuzzy logic technique in 1965. Recently this area has featured more in research. This fuzzy logic technique can be used for uncertainty and it can also be utilized in information granularity and imprecision. The generation of keen mapping among output and input space can be done effectively by means of fuzzy logic. Some of the main modules are given as follows:

The present classification table will be transferred onto a continuous classification in the initial stage. This operation in fuzzy logic is referred to as fuzzification. Then the domain experts will produce an inference engine based on the knowledge base. With the aid of this the operation will be carried out within the fuzzy domain. The database and rule base will come under the knowledge base. The generated fuzzy number will then be transferred back onto the single value called the "real world". This operation is referred to as defuzzification.

4.3.1 FUZZY NUMBER

Among the various existing quantities, fuzzy number is one among them. The value for this is imprecise. This value will be found to be similar to a normal single value number. The fuzzy number is defined as a membership function. The domain of the function is more specified. The domain contains real numbers in the range between the [0, 1] interval is referred to as a positive number. A specified value will be allocated for every numerical value in the domain. The highest possible value in this membership function is 1, similarly the lowest possible value is 0. The various forms for plotting fuzzy numbers are given as follows:

- Parabolic shaped fuzzy numbers.
- Trapezoidal fuzzy numbers.
- Triangular fuzzy numbers.

Fuzziness: Fuzzy numbers are referred to as distinct fuzzy sets which denotes the information is of uncertain quantity. They will be either exhibited as normal or convex and frequently denoted as single modal values. These fuzzy numbers are linked to some fuzziness or vagueness.

Mamdani-style inference: The fuzzy set in each rule will be correlated by means of an aggregation operator resulting in fuzzy inference. The gathered fuzzy set will be defuzzified further in order to obtain the outcome.

Membership function (MF): The degree to which the input fits to set or is otherwise correlated to the concept is referred to as function.

4.4   FIVE INPUTS FUZZY MODEL

In the proposed fuzzy model five kinds of input were provided. The five inputs are Performance, Security, Functionality, Portability, and Reusability. By means of these rules the process will be carried out in this model and effective estimation of selection effort will be achieved. The classification of input in fuzzy can be high, medium, and low. The outcome for the selection effort can be divided into very low, low, medium, high and very high. The five inputs model is shown in Figure 4.1

For fuzzifying the inputs, the subsequent membership functions are considered as high, medium. and low. They are shown in Figure 4.2.

Likewise, the selection of variables for output is considered, i.e., for Selection Effort five membership functions will be considered as shown in Figure 4.3.

4.5   FIVE INPUTS METHODOLOGY

All the inputs and outputs are fuzzified as shown in Figures 4.4 to 4.8.

The probability of every potential combination was taken into account leading to the formation of 3^5. In this 3^5 243 sets will be present. The output which is referred to as Selection Effort in this 243 set will be divided into low, medium, high, and very high. Based on this 243 set the fuzzy model is formed, and is represented below:

(1) The Selection Efforts will be low in cases of higher performance, security, functionality, portability, and reusability.
(2) The Selection Efforts will be low in cases of higher portability and reusability, and lower performance, functionality, and security,
(3) The Selection Efforts will be low in cases of higher security, functionality, portability, and reusability and lower performance.

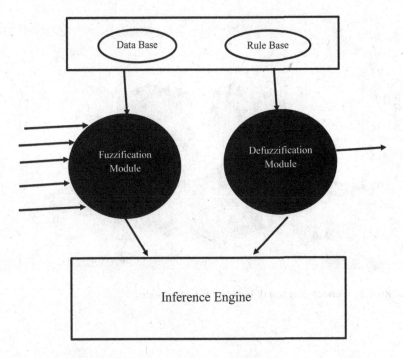

FIGURE 4.1 Five Inputs Fuzzy Model for Selection Efforts.

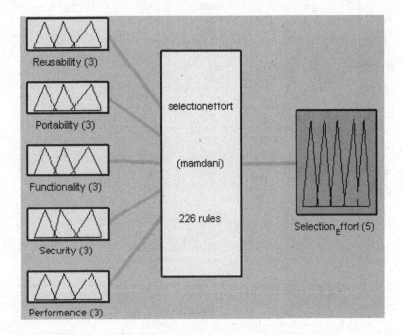

FIGURE 4.2 Inputs and outputs in Five Inputs Fuzzy model.

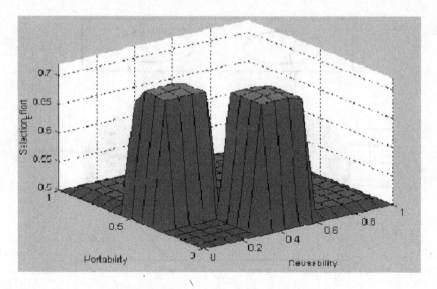

FIGURE 4.3   Surface diagram of Five Inputs Fuzzy model.

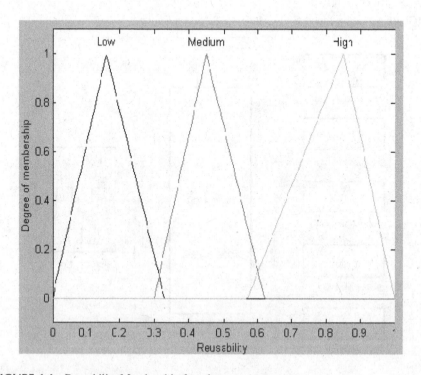

FIGURE 4.4   Reusability Membership function.

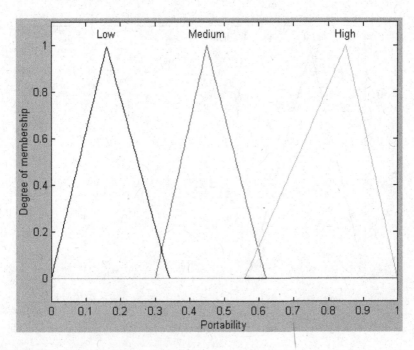

FIGURE 4.5 Portability Membership function.

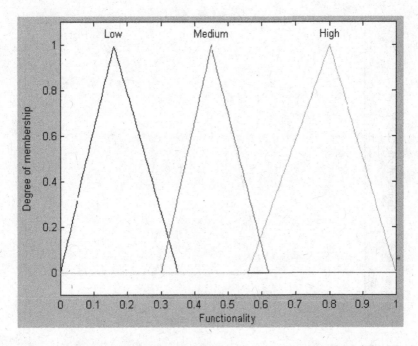

FIGURE 4.6 Functionality Membership function.

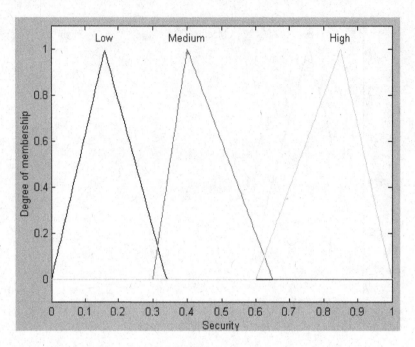

FIGURE 4.7   Security Membership function.

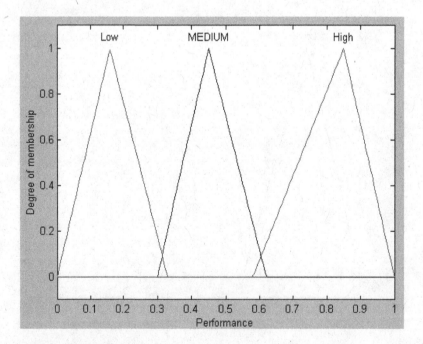

FIGURE 4.8   Performance Membership function.

(4) The Selection Efforts will be low in cases of higher security, functionality, and portability, lower performance, and medium reusability.

(5) The Selection Efforts will be low in cases of higher security, portability, and reusability, and lower performance and functionality.

(6) The Selection Efforts will be low in cases of higher performance, portability, and reusability, lower functionality and medium security.

(7) The Selection Efforts will be low in cases of higher security, portability, and reusability and lower performance and functionality.

(8) The Selection Efforts will be low in cases of higher portability and reusability, lower functionality and security and medium performance.

(9) The Selection Efforts will be low in cases of higher portability and performance, lower functionality and medium security and performance.

(10) The Selection Efforts will be high in case of higher reusability and lower performance, functionality, security, and portability.

(11) The Selection Efforts will be high in cases of higher reusability and functionality, lower performance and medium functionality, security, and portability.

In Table 4.2 the parameters and their values used in the Five Inputs Model are shown.

Table 4.3 shows the summary of five inputs used in the Five Inputs Model.

The results of Five Inputs model after running the model are presented in Table 4.4.

Selection efforts are calculated using a particular set of inputs using MATLAB® Fuzzy Toolbox. The Mamdani style of inference is used. The Fuzzy Rule Viewer is shown in Figure 4.9.

TABLE 4.2
Fuzzy System of Five Inputs Model

System		
S.No.	Parameter	Parameter Value
1	Name	"Selectioneffort"
2	Type	"mamdani"
3	Version	2.0
4	NumInputs	5
5	NumOutputs	1
6	NumRules	243
7	AndMethod	"min"
8	OrMethod	"max"
9	ImpMethod	"min"
10	AggMethod	"max"
11	DefuzzMethod	"centroid"

TABLE 4.3
Fuzzy Inputs for Five Inputs Model

Inputs	Name	Range	Number of MFs	MF1	MF2
Input 1	Reusability	[0,1]	3	"Low": trimf,[0 0.16 0.33]	"Medium": trimf,[03 0.45 0.62]
Input 2	Portability	[0,1]	3	"Low'": trimf,[0 0.16 0.34]	"Medium": trimf,[03 0.45 0.62]
Input 3	Functionality	[0,1]	3	"Low": trimf,[0 0.16 0.35]	"Medium": trimf,[03 0.45 0.62]
Input 4	Security	[0,1]	3	"Low": trimf,[0 0.16 0.34]	"Medium": trimf,[03 0.45 0.65]
Input 5	Performance	[0,1]	3	"Low": trimf,[0 0.16 0.33]	"Medium": trimf,[03 0.45 0.62]

TABLE 4.4
Fuzzy Output for Five Inputs Model

Output 1	
Name	Selection Effort
Range	[0 1]
Number of MF	5
MF1	"Very Low": "trimf", [0 0.12 0.23]
MF 2	"Low": "trimf", [0.2 0.32 0.42]
MF 3	"Medium": "trimf", [0.40 0.51 0.62]
MF 4	"High": "trimf", [0.60 0.75 0.82]
MF 5	"Very High": "trimf", [0.80 0.91 1.0]

4.6   EMPIRICAL EVALUATION

We calculate the value of the proposed factors for the component used in a Classroom-Based Project called the College Information System. This project has following three modules:

- Attendance system.
- Fees management system.
- Results system.

All these modules require the same Component Calendar which displays date, time, month, year, and some more facilities. To select these Components we visited two sites, www.componentsource.com and www.jars.com, and calculated the value of five factors as follows:

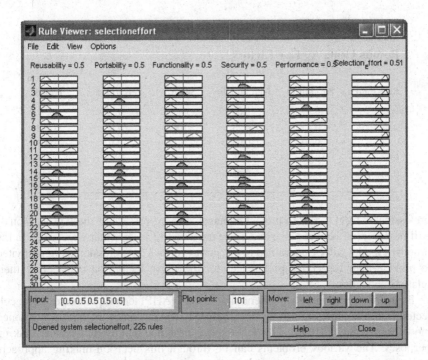

FIGURE 4.9 Selection Effort Rule Viewer.

Reusability: the value of reusability is given on the site. Each component that is on this site has stars attached to it. The higher the number of stars, the higher is the reusability.

Another method of calculating reusability is proposed by Sharma Arun (2009) and is based on the Artificial Neural Network (ANN)-based approach.

Portability: A component which is supported by most operating systems has higher portability.

Functionality: The functionality of a component can be calculated by the number of different tasks or facilities provided by the components.

Security: This factor can be calculated by seeing the security mechanism implemented on the particular component.

Performance: Performance can be calculated by the size of the component. The value of the factors is then supplied to the fuzzy rule base and the results are found as follows shown in Table 4.5.

4.7 WEIGHT ASSIGNMENT FACTORS FOR COMPONENT SELECTION EFFORTS

On the basis of this component-related technology some surveys were performed on software and professionally taught projects to evaluate the live analysis. In the case of teachers, the ranking will be from lecturer to professor and for software

TABLE 4.5
Results Using All Five Factors

Reusability	Portability	Functionality	Security	Performance	Selection Efforts
0.50	0.50	0.50	0.50	0.50	0.51
0.17	0.45	0.23	0.62	0.18	0.472
0.12	0.20	0.09	0.13	0.10	0.902
0.72	0.90	0.75	0.85	0.97	0.311
0.58	0.85	0.62	0.32	0.90	0.541

professionals from senior to project manager. The experience of these professionals will be different, between five and 12 years. The five factors that were given in the proposed model will be contained in the survey form. The professionals were invited to provide their respective preferences for these five factors on considering their related application.

By utilizing the Analytical Hierarchy Process (AHP) technique the response collected from the professionals will be analyzed. The design of the above-mentioned model was by Saaty (1994). This technique is designed mainly for decision-making purposes. The various problems can be through this decision-making approach. They can be used in evaluating choices in multi-objective decision situations, quantifying intangible factors, and decision makers in structuring complex decisions. For the set of elements, the related value can be obtained through this rational decision-making framework. This is why it is referred to as a powerful technology. In cases of comparison decision making, which is a tedious process, the AHP technique can also be used. This technique has numerous applications which include marketing, corporate planning, portfolio selection, and transportation planning. The application is mentioned as follows:

- The relative importance existing between the attributes can be developed by means of exhaustive paired comparison analysis or expert opinion.
- For every attribute, by means of certain algorithms the weight age will be developed.
- For every attribute, a similar analysis will be carried out for alternative solution strategies.
- In cases of every alternate strategy solution the single entire score will be developed.

In this survey 125 professionals were considered, the analysis was carried out in three phases and at least 25 persons were selected. In every phase the analysis was carried out with the help of the AHP technique and in the obtained result 8% variation was found. The operation of the data created was done using MS Excel. For every main parameter, the weight value is illustrated in Table 4.6. The values will be in the range of 0 to 1. The summation of the weighed values is will be 1.

TABLE 4.6
Weight Values Assignment for Each Factor

Factors	Reusability	Portability	Functionality	Security	Performance
Weights	0.21875	0.210938	0.175781	0.167969	0.226563

Correlation: Correlation is one of the most widely used statistical techniques. It is very useful in the fields of biology, economics, agriculture, psychology, etc. where there are certain relationships between pairs of variables. These relationships enable us to predict certain things. For example, an increase in production of a thing will raise the fall in price.

4.8 CORRELATION COEFFICIENT DEFINITION

The strength of linear association among two variables will be calculated. Most of the time the correlation will be between −1.0 and +1.0. The positive relationship will be obtained in cases of positive correlation. The negative relationship will be obtained in cases of negative correlation.

FORMULA

CORRELATION COEFFICIENT

$$\text{Correlation}(r) = N\Sigma XY - (\Sigma X)(\Sigma Y) / \text{Sqrt}\left(\left[N\Sigma X^2 - (\Sigma X)^2\right]\left[N\Sigma Y^2 - (\Sigma Y)^2\right]\right)$$

Where
ΣY^2 = Sum of square Second Scores.
ΣX^2 = Sum of square First Scores.
ΣY = Sum of Second Scores.
ΣX = Sum of First Scores.
ΣXY = Sum of the product of First and Second Scores.
Y = Second Score.
X = First Score.
N = Number of values or elements.

4.9 EMPIRICAL VALIDATION

Here we calculate the Component Selection Efforts by two methods: one is fuzzy logic-based and another is the Analytical Hierarchical Process (AHP). Finally, we calculate the correlation coefficient between these two outputs and the value of this comes out at 0.714. Hence, we conclude that the proposed technology is valid. Validation is shown in Table 4.7.

TABLE 4.7

Correlation between Selection Efforts Using Fuzzy Rule and Using AHP

Selection Efforts Using Fuzzy Rule Base (X)	Selection Efforts Using AHP (Y)	Correlation coefficient between X and Y
0.51	0.425	0.714
0.713	0.8160	
0.51	0.48	
0.51	0.3162	
0.572	0.475	
0.718	0.653	
0.713	0.572	
0.361	0.243	

EXERCISE

CASE STUDY 1

There is a food service provider, "Blue Pluto", which provides catering services for various occasions such as family functions, office parties, wedding parties, birthday parties, etc. It provides catering services either at the client's personal venue or in the venues with which there is a tie-up with Blue Pluto. The service offers different packages and also the client may customize based on their requirements. Catering is a multifaceted segment of the food service industry. Blue Pluto has many sites in different cities with some different services but presently they all work manually. Since Blue Pluto is a large venture it requires a software application to maintain master and transaction records and automate its system. Initially Blue Pluto wants to automate only one site. Along with automating basic functionalities, there are the following non-functional requirements with different weights too:

Reusability – Low.
Portability – High.
Functionality – Very High.
Security – High.
Performance – High.

Use the fuzzy-based effort estimation technique to estimate the effort required to develop the software

Reusability	Portability	Functionality	Security	Performance
0.65	0.5	0.8	0.45	0.4

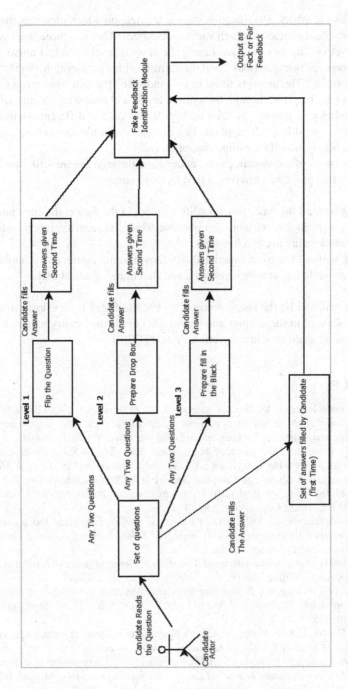

FIGURE 4.10 Fake Feedback Identification System (FFIS).

Case Study 2

There is a fake feedback identification system application which identifies the fakeness of feedbacks collected through various resources. This is a three-level system. The first level is "flip any two questions", the second level is "select answer from drop down box for two questions", and the third level is "preparing fill in the blank for two questions". The answers filled in by candidates the first time are compared with the answers received through these three levels and then with an internal computation feedback is marked as fake or fair. Based on the different environment systems can be modified and applied. For example, it could be used in colleges, universities, hotels, hospitals, online shopping, etc.

The framework of the system can be understood through Figure 4.10.

There are two proposals given by a development company.

A: Along with all the basic functionalities shown in the figure, the application will be very highly reusable and portable while the security, functionality, and functionality are at a low level.

B: Along with all the basic functionalities shown in the figure, in the application reusability, portability, security, and functionality are at high.

Which proposal will be the cheaper one if we use the fuzzy logic effort estimation approach discussed in the chapter and the weights of all the quality parameters are 0.25 for the Fake Feedback Identification System (FFIS)?

REFERENCES

Carrasco-Colomo-Palacios, R., López-Cuadrado, J. L., & Peñalvo, F. J. G. (2012). SEffEst: Effort estimation in software projects using fuzzy logic and neural networks. *International Journal of Computational Intelligence Systems*, 5(4), 679–699.

Martin, C.L., Pasquier, J.L., Yanez, C.M., Tornes, A.G. (2005). Software development effort estimation using fuzzy logic: A case study. *Proceedings of the sixth Mexican International Conference on Computer Science (ENC'05)*, Puebla, Mexico.

Parkash, K., Mittal, H., & Mittal, H. (2010). Software cost estimation using fuzzy logic. *ACM SIGSOFT Software Engineering*, 35(1), 1–7.

Patil, L. V., Shivale, N.M., Joshi, S.D., Khanna, V. (2014). Improving the accuracy of CBSD effort estimation using fuzzy logic. *IEEE International Advance Computing Conference, IACC*, Guragon, India.

Saaty, T. L. (1994). How to make a decision: The analytic hierarchy process. *Interfaces*, 24(6), 19–43. doi:10.1287/inte.24.6.19.

Seth, K., Sharma, A. Seth, A. (2009).Component selection efforts estimation–a fuzzy logic based approach. *International Journal of Computer Science and Security,(IJCSS)*, 3(3), 210–215.

Sharma, A., Grover, P. S., & Kumar, R. (2009). Reusability assessment for software components. *ACM SIGSOFT Software Engineering Notes*, 34(2):1–6.

Xu, Zhiwei, & Khoshgoftaar, Taghi M. (2004). Identification of fuzzy models of software cost estimation. *International Journal of Fuzzy Sets and Systems*, 145(1),141–163. Elsevier.

Zadeh, L. A. (1965). Fuzzy sets. *Information and Control*, 8(3), 338–353.

5 Estimating Component Integration Efforts
A Neural Network-Based Approach

5.1 INTRODUCTION

Integration is one of the most promising or important tasks for Component-Based Development (CBD). The Commercial off the Shelf (COTS) Component is selected to satisfy the system requirements and after selection these components should be integrated to form the required system. Component integration is vital to the Component Selection process, so the efforts invested in integrating these components must be calculated. This chapter gives a model for estimating the efforts invested in integrating these components. The CBD process can be divided into two parts: Component Selection and Component Integration. The total efforts invested in CBD can be estimated by adding these two efforts.

COTS Components are black box entities. An application may be developed by integrating these components. COTS products are in demand today, because developers believe that system quality is high and development time is low. COTS-Based Systems (CBS) have their own problems, e.g., selection of application component, integrating components, and maintenance. Due to all these advantages, companies are unable to find time and resources to reflect market-testing capabilities because CBS are expensive and difficult to manufacture, field, and support. There is an important reason for such difficulties in constructing these systems, as organizations ignore the point of proper integration of COTS components and assume that they work by throwing different components together. Their traditional engineering skills and processes can be thrown or drawn. These things do not constitute a COTS-based system.

A solution to these problems may be illustrated by answering the following types of questions:

(1) How we can find the most appropriate COTS product from the market?
(2) To select the most suitable software component according to our requirements we have to follow these three steps:

5.1.1 FORMULATION

(1) To establish a transparent and controlled process.
(2) To define the standard element for which vendors and products should be measured.
(3) To establish an evaluation model with weight measurements, and identify vendors and their products.

5.1.2 CONDUCT

Vendors need to be actively involved in the conduct phase – they must complete the questionnaire and participate in a conference room pilot (CRP) session. Once evaluation of the questionnaire in finished, qualified vendors should be invited to participate in the CRP. Set aside a few days for this process and make sure the product displays are relatively close. Develop a tool to capture the final vendor and product evaluation. To facilitate consideration, consider building one or more XY-graphs to describe each vendor/product's final evaluation and the relative position of each other. To facilitate observation, also consider using bar-diagrams to capture the assessment in each element. The spreadsheet should contain the details of a specific assessment and comments on the assessment to make it easier to remember the argument. Alternatively, the spreadsheet should have autonomy.

5.1.3 REPORT

The report phase consists of finalizing the report, the method used and the conclusions agreed in your group. Additional work with vendors may also be included in the report phase.

How to integrate these COTS products into the system?

COTS integration is all about binding different components together to form a system. In software engineering, the focus shifts from one of legacy system architecture and building to the requirement of considering system context (system attributes as schedule, requirements, support and operating set-ups, cost, etc.), better products and feasible specification of products.

5.2 PROBLEMS IN INTEGRATING COTS COMPONENTS

We know that there is a need to integrate COTS components with several different systems to get a new large complex system. In this section, problems which developers have to deal with in specifying requirements for integration are discussed. Not all the problems in the area are covered, but most serious problems are covered. Since it is an ongoing process the list of problems keeps on growing. Some common problems are as follows.

5.2.1 TO FIND DETAILS OF AVAILABLE PRODUCTS

These applications are very complex, it is very difficult to test these applications in a probing manner. Customers cannot set up the system themselves; it is the supplier's job to set up the systems. Suppliers would take couple of weeks to set them up.

One solution is for the customer to work together with the supplier to get to know their product, but it is not a good solution because this process takes a lot of time. Another drawback is that the customer's requirements are based on a particular system on which they worked closely, but this may unintentionally exclude other good suppliers. Also, customers do not know how to explain problems so that it remains supplier dependent.

Customers may read about the success of other companies' products, although, the products are very fast-paced and it gives a different image of what is available now. If this happens, then the customer has make a final selection which meets their requirements and selection criteria.

5.2.2 Not a Fixed Product Scope

The customer may not be sure about the size of product. At the initial phase a customer does not know about the platform, whether it is middleware, or it is already rolled out to customers. Choices for the same applications vary from customer to customer. Suppliers might provide the above-mentioned products or they may integrate a third-party product.

Many customers think that the entire system may be divided into modules and one can purchase modules. It may be seen easily from past experience that both applications and middleware are so closely connected that it is not easy to port from one middleware to another although they claim to strictly follow open standards, e.g., CORBA and J2EE (Gorton and Liu, 2002, 2003). The results in a situation where COTS products are not available. Choosing and buying middleware first is a good strategy if the rest are developed from scratch. If the rest are COTS products, then this process is not going to work properly.

5.2.3 Late Maintenance of Highly Complex Areas

After signing the contract, the supplier has to deliver, as discussed earlier. It may be the case that glue codes were developed by the supplier long ago, hence the supplier can extend the system, and the customer will use that system. This is theoretical. However, often it is seen that supplier is unable to fulfill its promise.

Common difficulty areas are performance (response time) and integration with other products. For example, the supplier's current installations have about 30 customers and work well with them. But when faced with a planned 2000 customers, the response time skyrockets. Or, the connection to System X seemed easy, but in practice it did not work as expected. Most customers did not care about this before. What is the problem, they ask? If the supplier does not deliver as promised, he does not get his money; he also has to pay a fine. These customers think the supplier is just lazy and the money will cure him. Any problems may not be solved by the supplier. Such projects can drag on for years. The customer never gets the appropriate system and the supplier loses money.

The main issue faced by parties is the high chances of being late. The supplier is inclined to deliver the easy part first while delaying more difficult one. Ideally, it is the customer's responsibility is to demand a guarantee for complex areas before selecting a component among those available. However, it is very costly and not realistic.

Proponents can do so during the proposal.

One can wonder now – what are the factors that satisfy the requirements for proper integration? The answer is –

for harmonious CBS proper integration is needed. The system does not work correctly or is incomplete without the right components.

What are the necessary major factors for maintenance of the systems implemented outside?

Software management is known as one of the most expensive resources in the software development process. Very soon, the development phase of software will be replaced by integration in CBD.

There is a need for maintenance at every phase of software. Hence, software developers need support for every sustainability perspective, and for that support, they need to understand sustainability and its impact on CBS. In particular, software consistency with COTS is different from legacy software management, with the possibility of managing operations other than by the developer. Whether this component is an "internally" implemented one or is purely outsourced.

The importance of discussing COTS selection and integration show up when considering that COTS products are developed to be generic, however they can be integrated into a system and are used in a specific context with certain dependencies. COTS components may be heterogeneous in nature, for example, they may have different interfaces, support different business protocols, and use different data formats and semantics. Hence, these mismatches should be eliminated before integrating COTS components into a CBS.

Differences between COTS products and system integration exist due to their architectural differences and constraints. These mismatches should be removed at the time of integration. Thus, a classification of mismatches or incompatibilities can be useful for COTS integration.

There is a significant amount of effort that is invested in developing a system by integrating these products. There are so many existing approaches for calculating the efforts invested in integrating the components.

Nowadays there are many studies using neural networks for predicting software quality matrices. Heiat (2002) compares prediction performance of neural networks and radial basis function neural networks through regression analysis. This method shows that neural network methods improve mean absolute percentage errors as compared to regression analysis.

Reddy et al. (2008) combined multilayer perceptron networks with COCOMO II for software cost estimation. The proposed method is evaluated on 63 COCOMO II projects. The main aim of this proposed method is to train multilayer perceptron in better way. In this proposed model, in intermediate layer efforts factors and scale factors are detailed and these factors are evaluated by COCOMO II. The Sigmoid Function was used as the activation function for the intermediate layer. An algorithm is used for teaching backpropagation. The data training phase is 80% and testing phases is 60%. The results suggest that the standard error of relative error in the composition model is less than the COCOMO II model.

We proposed another neural network-based approach. In this approach we define three factors which affect the Component Integration Efforts to a great extent. These factors are:

(1) Interaction Complexity.
(2) Understanding.
(3) Component Quality.

5.3   FACTORS AFFECTING COMPONENT INTEGRATION EFFORTS

5.3.1   INTERACTION COMPLEXITY

Components are pure black box entities. No-one can check the source code of components. Interaction with these components is possible only through interfaces. Hence, the interface acts as the main source to understand, implement, maintain, and reuse components. These interfaces explain individual elements of a component systematically.

Hence, complexity of these components is very crucial for correctly estimating overall component complexity.

5.3.2   UNDERSTANDING

The understanding of a component may be defined as the familiarity of the component with the environment. The environment can be both software and hardware. So, if a component is very familiar with the environment then integration efforts will be fewer, otherwise it will be high.

5.3.3   COMPONENT QUALITY

By Component Quality we mean that we select a well-suited component for our application. If the selection of a component is right, then the integration efforts will be low.

5.4   ARTIFICIAL NEURAL NETWORK-BASED APPROACH

In this chapter we have discussed the Artificial Neural Network (ANN) technique, measuring the integration efforts of a software component. ANNs are impressive techniques in the area of clustering and classification [MAINT 1, MAINT 2]. The following are the reasons for selecting Neural Networks:

• An ANN is adaptive and can classify patterns easily.
• The complexity of a network is adjusted by an ANN to that of a problem, hence, it produces better outcomes as compared to other analytical models.

- Neural Network techniques are known as "black boxes". Users cannot understand the process easily.
- There are no fixed guidelines for considering neural networks.

5.5 NEURAL NETWORK ARCHITECTURE

Figure 5.1 describes an elementary neuron with *n* input. A weight value *w* is assigned to each input. Transfer function *f* gets its input from the sum of all weighted inputs. For output generation, neurons may use any differentiable function.

5.6 MATLAB® NEURAL NETWORK TOOLBOX

MATLAB Neural Network Toolbox consists of many functions and utilities. It discusses how to use these functions for the creation and training of networks. One can also simulate and visualize neural networks in respect of verification and validation. Once the data for network training is available for analysis, utilities like interpolation, statistical analysis, equation solvers, and optimization routines can be used to plot the training error functions, monitor the change in weight matrix, and obtain the real-time network outputs to verify their accuracy. Transfer functions in the toolbox include hard limit, symmetric hard limit, log sigmoid, linear, saturated linear, and tan sigmoid functions. There are two important algorithms in neural networks:

- Levenberg–Marquardt.
- Bayesian Regularization.

5.7 EXPERIMENTAL DESIGN

The aim of this work is whether ANN can be used as tool for measuring the Integration Efforts of Component Based Systems. In this section, Integration Efforts are a number of factors mentioned above. Integration Efforts may be measured using these factors. These factors are considered as input min–max normalization is used.

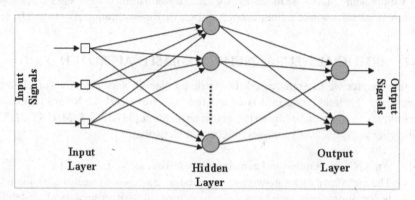

FIGURE 5.1 General architecture of Artificial Neural Network.

Min–Max normalization is a linear transformation on the original data. The original input range is transferred into a new one (generally, 0–1).

Let min$A$ = minimum value for attribute A.

max$A$ = maximum value for attribute A.

let value of $A = p$, it is mapped into a new value p'.

$$p' = \frac{p - \min A}{\max A - \min A}$$

The input factors are classified as low, medium, and high categories. Rules have been designed based on various combinations of input to predict the output. In total there will be 33 rules.

The network is trained using the trainlm function of the feed forward propagation algorithm. The Windrow–Hoff learning rule may be generalized using backpropagation. For this purpose, the transfer function used to be non-linear. Input vectors and the corresponding target vectors are used to train a network until it can approximate a function, associate input vectors with specific output vectors, or classify input vectors in an appropriate way as defined by you. Networks with biases, a sigmoid layer, and a linear output layer are capable of approximating any function with a finite number of discontinuities.

Accurately trained back propagation networks provide adequate answers for the inputs they have never processed. In general, the input of the vectors used in the new input training is the same as the output of the new input. This generalization makes it possible to train the network in a representative group of asset pairs and get good results without training the network on all possible input/output pairs.

Trainlm is a network training function that updates weight and bias values according to Levenberg–Marquardt optimization. Trainlm is often the fastest propagation algorithm in the toolbox and is highly recommended as a first-choice supervised algorithm, although it does require more memory than other algorithms.

Here, Tansig (a linear transfer function) is used as shown in Figure 5.2. It calculates a layer's output from its input.

At first, a training set is made up of 120 examples. The training examples have 35, 35, and 45 cases of, low, medium, and high categories respectively. The training exemplars were chosen randomly. The initial set was again extended to 135 examples having 40, 45, and 40 cases respectively, so the effect of the increment may be noticed for network performance. This small increment in training examples from

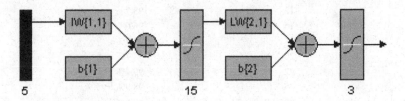

FIGURE 5.2 Tansig Function.

120 to 135 as compared to 243 possible inputs did not contribute significantly to performance.

ANN considers this task as the pattern recognition problem in which patterns are aligned to give number of classes. Through a training session neural networks recognize to pattern of given data. In a training session a network is supplied with a set of inputs and their category. After that a new pattern that has not been supplied earlier is passed through the network, but it is from the same pattern class by which the network was trained. The network is able to identify the class of that pattern because of the information it has extracted from the training data.

5.8   RESULTS

Training of networks is done by adjusting the neuron weight in the hidden layer to minimize the difference between the two outputs (actual, desired). The network learns by finding a vector of connection weights that minimizes the desired error on the training data set. To conduct this experiment networks are ensemble. At first these networks have five neurons and later on, the number of neurons may be increased up to 40. Among all trained networks the network with the best performance is the most suitable. By repeating the process, it is found that network is best trained by taking 22 neurons in the hidden layer. The accuracy of trained networks at 22 neurons is found to be 99.4%. Results with several different numbers of neurons are shown in Table 5.1.

CASE STUDY

Let's write an integration assessment for a Patient Data Recording system.

As the name suggests, using this system records various information about patients, such as diseases and their treatment information. Also, this system keeps a record of why particular treatment is given and what in future can be done to treat patients. There must be an option for transferring data digitally. Traditional patient

TABLE 5.1
Training Data Accuracy

Neurons in the Hidden layer	Accuracy
5	89.1%
10	93.1%
12	93.7%
15	96.0%
18	97.7%
20	97.7%
22	99.4%
25	98.9%
30	93.7%

records are more textual. Structural data such as measurement and diagnosis are ingrained in textual data and because of this it is difficult to use for an overview of a patient's condition or statistics. The system should be implemented in such a way that it helps hospital staff to record data in a better manner and can retrieve data in several ways. The major issue is that there is a number of services which provide different information about patients when asked, for example, pathology tests, medicine requirements, ultrasound records, X-ray information, food suggestions, and many more. All of these service providers have their own data format. In computer science we make classes and subclasses, in addition, most of the services are automatically managed through specialized subsystems (production). Some of the tests performed in laboratories are fully automated. Samples with codes are included in the its class, and test findings are obtained in the database shortly afterwards. Databases and the functionality of Data Recording Systems should be integrated with a production system score. These systems do not even have a common communication protocol.

REFERENCES

Gorton, I., & Liu, A. (2002). Streamlining the acquisition process for large-scale COTS middleware components. In: J. Dean & A. Gravel (Eds.), *International Conference on COTS-Based Software Systems, ICCBSS 2002, LNCS, 2255* (pp. 122–131), vol 2255. Springer, Berlin, Heidelberg

Heiat, A. (2002). Comparison of artificial neural network and regression models for estimating software development effort. *Information and Software Technology, 44*(15), 911–922.

Reddy, S., Rao, P. S., Raju, K., & Kumari, V. V. (2008). A new approach for estimating software effort using RBFN network. *International Journal of Computer Science and Network Security, 8*(7), 237–241.

Appendix A: Data Tables Used for Use Cases

TABLE A.1
Complexity Level Evaluation for CP1

	0–4 NEM	5–8 NEM	9–12 NEM	≥13 NEM
0–1 NSR	Low	Low	Average	High
2–3 NSR	Low	Average	High	High
4–5 NSR	Average	High	High	Very High
>5 NSR	High	High	Very High	Very High

TABLE A.2
Complexity Level Evaluation for CP2

0–2 NSR	0–5 NOA	6–9 NOA	10–14 NOA	≥15 NOA
0–4 NEM	Low	Low	Average	High
5–8 NEM	Low	Average	High	High
9–12 NEM	Average	High	High	Very High
≥13 NEM	High	High	Very High	Very High

(a)

3–4 NSR	0–4 NOA	5–8 NOA	9–13 NOA	≥14 NOA
0–3 NEM	Low	Low	Average	High
4–7 NEM	Low	Average	High	High
8–11 NEM	Average	High	High	Very High
≥12 NEM	High	High	Very High	Very High

(b)

≥5 NSR	0–3 NOA	4–7 NOA	8–12 NOA	≥13 NOA
0–2 NEM	Low	Low	Average	High
3–6 NEM	Low	Average	High	High
7–10 NEM	Average	High	High	Very High
≥11 NEM	High	High	Very High	Very High

(c)

System Component Type	Description	Complexity			
		Low	Average	High	Very High
PDT	Problem Domain Type	3	6	10	15
HIT	Human Interface Type	4	7	12	19
DMT	Data Management Type	5	8	13	20
TMT	Task Management Type	4	6	9	13

TABLE A.3
Degree of Influences of Twenty Four General System Characteristics

ID	System characteristics	DI	ID	System Characteristics	DI
C1	Data Communication	.....	C13	Multiple Sites	.....
C2	Distributed Functions	.....	C14	Facilitation of Change	.....
C3	Performance	.....	C15	User Adaptively	.....
C4	Heavily Used configuration	.....	C16	Rapid Prototyping	.....
C5	Transaction Rate	.....	C17	Multiuser Interactivity	.....
C6	Online Data Entry	.....	C18	Multiple Interface	.....
C7	End User Efficiency	.....	C19	Management Efficiency	.....
C8	Online Update	.....	C20	Developer's Professional Competence	.....
C9	Complex Processing	.....	C21	Security	.....
C10	Reusability	.....	C22	Reliability	.....
C11	Installation Ease	.....	C23	Maintainability	.....
C12	Operational Ease	.....	C24	Portability	.....
TDI	Total Degree of Influence (TDI)				

Appendix B: Review Questions

1. Among all the proposed metrics list any five metrics which, in your opinion, predict component development effort most precisely.
2. Which of these proposed metrics significantly adds to the predictive ability of COCOMO?
3. Can the proposed metrics be used as a basis for further remuneration options for COCOMO?
4. Describe the top model/technique for software cost estimation.
5. Case study a safety critical system and find out the differences in estimating software costs compared to normal systems.
6. How do we evaluate the effectiveness of software effort estimation models?
7. Explain in detail the importance of "Project Methodology" as a parameter for software cost estimation.
8. Define scalability as far as the software cost estimation is concerned.
9. What is the importance of the use case points method in software effort estimation?
10. What are the most used approaches for evaluation of software effort estimation models?
11. Is it possible to do a software cost estimation before requirement collection?
12. What method can be used to measure the accuracy of prediction for an effort estimation technique in software development?
13. Explain the research gaps in the deep learning, machine learning ... soft computing techniques in the analysis of estimating efforts.
14. List the available simulation tools to work on soft computing techniques.
15. What is the state of the knowledge surrounding COTS technology implementation?
16. Describe the forerunners of both obstacles and designers of COTS technology implementation.
17. What are the issues for successful adoption and performance of COTS technology?

Recent Trends

In recent years, there has been remarkable development in the software and research industry. Initially, deterministic techniques were used to solve Non-deterministic Polynomial-time NP-hard problems and complex systems. Nowadays, evolutionary algorithms are used to frequently solve complex problems. These algorithms are based on Darwin's theory of evolution. Chromosomes, the key element in evolutionary algorithms, are used to represent individuals. These chromosomes are fixed-length strings. All these algorithms work in two phases – firstly, a random selection of the individual population is made and secondly, a fitness function is used for selecting the most suitable candidate. Crossover and mutation operations are used to find new generations. It may be stated that evolutionary algorithms use iterative progress, such as population growth. Population selection is made using a guided random search and by parallel processing to find the desired results. All these processes are inspired by the biological evolution mechanism. This evolution has many applications in computer science.

The following are the major evolutionary algorithms: Genetic algorithm, Honeybee Ant Colony optimization, Particle Swarm optimization, Differential Evolution Harmony search, Genetic Programming Cultural algorithm, and Cuckoo search.

One of the emerging research areas is to find a Genetic algorithm contribution to software engineering. Software engineering implies a systematic approach to maintain the system. Using software metrics, we can measure the performance of the software. The main aim of software testing is to find errors or "bugs" present in the software. Software quality assurance is used to judge and improve its quality. Many authors used different techniques to estimate the cost of the software module. Recent advances in the area of component-based software cost estimation are using genetic algorithms to find the software cost.

Exponential growth had been noted in the use of genetic algorithms in recent years. Genetic algorithms are used in heterogeneous fields, e.g., Agriculture, Physics, Mechanical Engineering, Chemistry, Astronomy, Computer Science, Medical Science, etc. The accuracy of the genetic algorithm is excellent.

The following are the research that has been done in this area:

(1) Maleki and Ebrahimi (2014) proposed a method that is a combination of the Firefly algorithm and Genetic algorithm. The NASA 93 dataset is used for the evaluation. In this method, the elitism operation of a Genetic algorithm is used to find the most suitable answer for factors and fitness function of assessment, and this results in a lower error rate. The findings indicate that the average value of the relative error predicted by the COCOMO model is 58.80, and in Genetic and Firefly algorithms is 38.31 and 30.34, respectively, and in the proposed composition model is 22.53.

(2) Sharma and Fotedar (2014) proposed a review of different datamining techniques that are useful for effort estimation. One of the important tasks in cost prediction is effort estimation and it is listed under the planning phase of software project management. Some of the methods taken into account are clustered techniques, for example, K-Means, K-NN-K-Nearest Neighbor, Regression technique, Multivariate Analysis for Regression Splines (MARS), Ordinary Least Square regression (OLS), Support Vector Regression (SVR), Classification And Regression Trees (CART); and classification techniques, namely Support Vector Machine (SVM), and Case Based Reasoning (CBR). Hybrid approaches can be used for the above methods for enhancing effort estimation. The proposed technique uses some of the data mining methods that have been detailed to improve the precision of software effort estimation. Effort is being calculated on the basis of the MMRE rate. The lower the MMRE value the technique is supposed to be better. From now onwards, a hybrid approach for any of the datamining methods for increasing the precision in effort estimation can be used. Some of the datasets that can be used are National Aeronautics and Space Administration (NASA), COCOMO 81, International Function Point User Group (IFPUG). Researchers can also use datasets from the COTS project.

(3) Morera et al. (2017) validated a genetic framework for effort estimations. The authors also performed a sensitivity analysis for different genetic configurations. Performance results were investigated for the best-learning algorithms. There is much scope for research into the learning schemes used, including 600 learning schemes, eight different processors, five attribute selectors, and 15 modeling techniques. The elitism genetic framework technique is used to automatically select the best learning scheme. A learning scheme is called the best if it has the highest coefficient correlation together with data processing, attribute selection, and learning algorithms. These selected learning schemes are applied to datasets extracted from the ISBSG R12 dataset.

The results show that the performance of the proposed Genetic algorithm is equally good when compared with an exhaustive framework. Sensitivity analysis shows the stability among different genetic configurations. The proposed framework is stable, and performance is better than a random approach. Results show that assembling machine-learning techniques can further optimize efforts estimation.

(4) Krishna and Krishna (2012) proposed a new technique for estimating software costs using Particle Swarm optimization and Fuzzy Logic. Software cost estimation depends upon the size of projects and its proposed parameters. Uncertainty is measured using Fuzzy Logic and Particle Swarm optimization is used for parameters. In these proposed techniques, the triangular membership function of Fuzzy Logic is used. Authors evaluated the proposed method on NASA 63 dataset. Findings show that the proposed technique has a lower rate of error compared to other models.

(5) Dizaji and Gharehchopogh (2015) used a hybrid of Ant Colony optimization and Chaos optimization for estimating software cost. To generate random data Lorenz mapping is used for the Chaos optimization algorithm and training is done by using the Ant Colony algorithm. This proposed algorithm is evaluated using NASA 63 dataset. Better performance, as compared to the COCOMO model, is noted in this algorithm in terms of performance and errors.

(6) Gharehchopogh and Pourali (2015) scrutinized adjusted COCOMO II model by using a dataset of NASA projects to examine the effect of the developed model and showed the effectiveness of the proposed algorithm in the correction of parameters of COCOMO II. Experiment results show that this model offers an excellent estimate for software cost estimation. In this book, the authors aim to develop an evolutionary model for software cost estimation by using continuous genetic algorithms.

At the onset, a lot of techniques are being used for estimating efforts of CBS using genetic algorithms. Some of these techniques use hybrid approaches with Evolution algorithms, Datamining, Machine Learning, Fuzzy Logic, etc.

The trend of future research is to use genetic algorithms for software effort estimations.

REFERENCES

Dizaji, Z. A., & Gharehchopogh, F. S. (2015). A hybrid of ant colony optimization and chaos optimization algorithms approach for software cost estimation. *Indian Journal of Science and Technology, 8*(2), 128–133.

Gharehchopogh, F. S., & Pourali, A. (2015). A new approach based on continuous genetic algorithm in software cost estimation. *Journal of Scientific Research and Development, 2*, 87–94.

Krishna, B., & Krishna, T. K. R. (2012). Fuzzy and swarm intelligence for software effort estimation. *Advances in Information Technology and Management, 2*(1), 246–250.

Maleki, L., & Ebrahimi, F. S. (2014). Gharehchopogh, "A hybrid approach of firefly and genetic algorithms in software cost estimation". *MAGNT Research Report, 2*(6), 372–388.

Murillo-Morera, J., Quesada-López, C., Castro-Herrera, C., & Jenkins, M. (2017). A genetic algorithm-based framework for software effort prediction. *Journal of Software Engineering Research and Development, 5*(1), 4. doi:10.1186/s40411-017-0037-x.

Sharma, M., & Fotedar, N. (2014). Software effort estimation with data mining techniques-A review. *International Journal of Engineering Sciences and Research Technology,* pp [1646-1653 3*.

RECENT RESEARCH QUESTIONS IN THE SAME FIELD

RESEARCH QUESTION 1

Among all the genetic frameworks find out the best suitable framework for performance as compared to the exhaustive baseline framework.

By framework, we mean generation, population mutation levels, and crossover.

RESEARCH QUESTION 2

Find out similarities and differences in the genetic framework for evaluation and prediction phases.

RESEARCH QUESTION 3

List the genetic framework learning scheme that is most often selected.

RESEARCH QUESTION 4

Find out the best-performing learning scheme reported in terms of evaluation criteria.

Index

Printed in the United States
By Bookmasters